Information and

This shop manual contains several sections each covering a specific group of wheel type tractors. The Tab Index on the preceding page can be used to locate the section pertaining to each group of tractors. Each section contains the necessary specifications and the brief but terse procedural data needed by a mechanic when repairing a tractor on which he has had no previous actual experience.

Within each section, the material is arranged in a systematic order beginning with an index which is followed immediately by a Table of Condensed Service Specifications. These specifications include dimensions, fits, clearances and timing instructions. Next in order of arrangement is the procedures paragraphs.

In the procedures paragraphs, the order of presentation starts with the front axle system and steering and proceeding toward the rear axle. The last paragraphs are devoted to the power take-off and power lift sys-te... ...ular specifications pertaining to wear limits, torquing, etc.

HOW TO USE THE INDEX

Suppose you want to know the procedure for R&R (remove and reinstall) of the engine camshaft. Your first step is to look in the index under the main heading of ENGINE until you find the entry "Camshaft." Now read to the right where under the column covering the tractor you are repairing, you will find a number which indicates the beginning paragraph pertaining to the camshaft. To locate this wanted paragraph in the manual, turn the pages until the running index appearing on the top outside corner of each page contains the number you are seeking. In this paragraph you will find the information concerning the removal of the camshaft.

More information available at Clymer.com
Phone: 805-498-6703

Haynes Publishing Group
Sparkford Nr Yeovil
Somerset BA22 7JJ England

Haynes North America, Inc
861 Lawrence Drive
Newbury Park
California 91320 USA

ISBN 10: 0-87288-561-5
ISBN-13: 978-0-87288-561-5

© **Haynes North America, Inc. 1954**
With permission from J.H. Haynes & Co. Ltd.

Clymer is a registered trademark of Haynes North America, Inc.

Printed in the U.S.A.

Cover art by Sean Keenan

All rights reserved. No part of this book may be reproduced or transmitted in any form or by any means, electronic or mechanical, including photocopying, recording or by any information storage or retrieval system, without permission in writing from the copyright holder.

While every attempt is made to ensure that the information in this manual is correct, no liability can be accepted by the authors or publishers for loss, damage or injury caused by any errors in, or omissions from, the information given.

SHOP MANUAL

COCKSHUTT

MODELS 20-30-40-50

CO-OP

MODELS E2-E3-E4-E5

GAMBLE'S FARMCREST

MODEL 30

Co-Op model E2 is the same as Cockshutt model 20. Co-Op model E3 and Gamble's Farmcrest model 30 are the same as Cockshutt model 30. Co-Op model E4 is the same as Cockshutt model 40 and Co-Op model E5 is the same as Cockshutt model 50. All subsequent data in this manual will be listed under the Cockshutt model numbers only. When working on Co-Op or Gamble's Farmcrest tractors, refer to the equivalent Cockshutt model numbers.

LOCATION OF SERIAL NUMBERS

ENGINE. Engine serial number is stamped on left side of cylinder block.

TRACTOR. Tractor serial number is stamped on left side of main frame.

IDENTIFICATION

	20	30	40	50
Single Wheel Tricycle	No	Yes	Yes	No
Dual Wheel Tricycle	Yes	Yes	Yes	Yes
Adjustable Axle	Yes	Yes	Yes	Yes
Non-Adjustable Axle	No	Yes	Yes	Yes

(For tractor cross-sectional views, refer to pages CS-20 and CS-92)

INDEX (By Starting Paragraph)

	20	Non-Diesel 30	Diesel 30	Non-Diesel 40	Diesel 40	Non-Diesel 50	Diesel 50
BELT PULLEY	320	325	325	327	327	327	327
BRAKES	300	303	303	307	307	310	310
CARBURETOR (NOT LP-GAS)	100	100	...	100	...	100	...
CARBURETOR (LP-GAS)	...	105	...	105	...	...	...
CLUTCH	210	210	210	210	210	210	210
COOLING SYSTEM							
Radiator	195	195	195	195	195	195	195
Thermostat	196	196	196	196	196	196	196
Water pump	197	199	199	199	199	199	199
DIESEL FUEL SYSTEM							
Energy cells	...	...	175	...	175	...	175
Fuel filters	...	...	132	...	132	...	132
Injection pump (APE)	...	...	152	...	152	...	152
Injection pump (PSB)	...	...	160	...	160	...	160
Nozzles	...	...	133	...	133	...	133
Preheater	...	...	176	...	176	...	176
Quick checks	...	...	130	...	130	...	130
DIFFERENTIAL	275	277	277	282	282	282	282
ENGINE							
Cam followers	52	53	53	53	53	53	53
Camshaft	61	63	63	63	63	63	63
Connecting rods & bearings	71	71	71	71	71	71	71
Crankshaft	72	73	73	73	73	73	73
Cylinder head	43	44	45	44	45	44	45
Engine removal	40	41	41	42	42	42	42
Flywheel	77	77	77	77	77	77	77
Ignition timing	201	201	...	201	...	201	...
Injection timing (APE)	...	...	152	...	152	...	152
Injection timing (PSB)	...	...	160	...	160	...	160
Main bearings	72	73	73	73	73	73	73
Oil pump	78	78	78	78	78	78	78
Pistons & rings	66	67	68	67	68	67	68
Piston pins	70	70	70	70	70	70	70
Piston removal	65	65	65	65	65	65	65
Rear oil seal	74	75	75	75	75	75	75
Rocker arms	...	54	54	54	54	54	54
Timing gear cover	56	57	57	57	57	57	57
Timing gears	58	60	60	60	60	60	60
Valves & seats	46	47	47	47	47	47	47
Valve guides & springs	48	50	50	50	50	50	50
Valve timing	55	55	55	55	55	55	55
FINAL DRIVE							
Axle shaft	287	290	290	296	296	296	296
Bull gears	287	290	290	296	296	296	296
Bull pinions	286	293	293	295	295	295	295
FRONT SYSTEM							
Axle type	7	7	7	7	7	7	7
Tricycle type	1	2	2	2	2	3	3
GOVERNOR (NON-DIESEL)	180	181	...	181	...	181	...
Diesel (APE)	...	...	157	...	157	...	157
Diesel (PSB)	...	...	166	...	166	...	166
HYDRAULIC SYSTEM							
Adjustments	366	366	366	366	366	366	366
Control valves	375	375	375	375	375	375	375
Lubrication	365	365	365	365	365	365	365
Pump	380	381	383	385	385	385	385
Work cylinders	390	390	390	390	390	390	390
LP-GAS SYSTEM							
Carburetor	...	105	...	105	...	...	...
Converter	...	112	...	112	...	...	...
Trouble shooting	...	116	...	116	...	...	...
POWER TAKE-OFF (Live type)	...	332	332	350	350	350	350
Non-continuous type	320	330	330	...	...	...	...
REAR AXLE	287	290	290	296	296	296	296
STEERING GEAR	20	25	25	32	32	32	32
TRANSMISSION							
Basic procedures	231	244	244	259	259	259	259
Major overhaul	236	250	250	265	265	265	265
Remove & reinstall	230	...	...	258	258	258	258

Sectional Views

COCKSHUTT 20-30-40-50

Cockshutt Model 20

Cockshutt Model 30

COCKSHUTT 20-30-40-50

Condensed Service Data

Tractor Models	20	Non-Diesel 30	Diesel 30	Non-Diesel 40	Diesel 40	Non-Diesel 50	Diesel 50
GENERAL							
Engine Make	Continental	Buda	Buda	Buda	Buda	Buda	Buda
Engine Model	*F-140	4B153	4BD153	6B230	6BD230	6B273	6DA273
No. Cylinders	4	4	4	6	6	6	6
Bore—Inches	*3 3/16	3 7/16	3 7/16	3 7/16	3 7/16	3 3/4	3 3/4
Stroke—Inches	*4 3/8	4 1/8	4 1/8	4 1/8	4 1/8	4 1/8	4 1/8
Displacement—Cubic Inches	*140	153	153	230	230	273	273
Compression Ratio—Gasoline	6.75:1	6.18:1		6.18:1		6.6:1	
Compression Ratio—Distillate	5.03:1	4.7:1		4.7:1			
Compression Ratio—Diesel			15:1		15:1		14.3:1
Compression Ratio—LP-Gas		7.12:1		7.12:1			
Pistons Removed From?	Above	Above	Above	Above	Above	Above	Above
Main Bearings, Number of	3	3	3	7	7	7	7
Main and Rod Bearings Adjustable?	No	No	No	No	No	No	No
Cylinder Sleeves	None	Wet	Wet	Wet	Wet	None	None
Forward Speeds, Std.	4	4	4	6	6	6	6
Forward Speeds, Opt.	None	8	8	None	None	None	None
Electrical System Voltage	6	6	12	6	12	6	12
Generator & Starter Make	A-L	A-L	A-L	A-L	A-L	A-L	A-L
TUNE-UP							
Firing Order	1-3-4-2	1-3-4-2	1-3-4-2	1-5-3-6-2-4	1-5-3-6-2-4	1-5-3-6-2-4	1-5-3-6-2-4
Valve Tappet Gap	.014H	.012H	.012H	.012H	.012H	.012H	.012H
Inlet Valve Seat Angle	30°	45°	45°	45°	45°	45°	45°
Exhaust Valve Seat Angle	45°	45°	45°	45°	45°	45°	45°
Ignition Distributor Make	A-L	A-L		A-L		A-L	
Ignition Distributor Model	IAD6004-1C	IAD6003-1A		IAD6001-1B		IAD6001-1B	
Breaker Contact Gap	.020	.020		.020		.020	
Ignition Timing-Retard (Gasoline)	6°BTC	10° BTC		TDC		TDC	
Ignition Timing—Distillate & LPG	TDC	10° BTC		TDC			
Injection Timing			See Paragraph 152 or 160				
Flywheel Mark Indicating							
Retard Timing (Gasoline)	6° BDC	Static Spark		TDC		TDC	
Retard Timing (Distillate & LPG)	DC	Static Spark		TDC		TDC	
Injection Timing			FPI		FPI		FPI
Spark Plug Make	Champion	Champion		Champion		Champion	
Model	8 Com.	J11		J11		J11	
Electrode Gap	.025	.025		.025		.025	
Carburetor Make	M.S	M.S or Zen.		Zen.		Zen.	
Model	TSX	TSX or 161		162		162	
Float Setting	1/4	1/4 or 1 5/32		1 39/64		1 39/64	
Calibration			See Standard Units Section				
Engine No Load RPM	2000	1810	1820	1850	1820	1850	1820
Engine Loaded RPM	1800	1650	1650	1650	1650	1650	1650
PTO Loaded RPM	563	545	545	530	530	530	530
BP Loaded RPM	1160	1336	1336	1000	1000	1000	1000
SIZES—CAPACITIES—CLEARANCES							
(Clearances in thousandths)							
Crankshaft Journal Diameter	2.2495	2.4975	2.4975	2.4975	2.4975	2.4975	2.4975
Crankpin Diameter	1.937	1.998	1.998	1.998	1.998	1.998	1.998
Camshaft Journal Diameter, Front (No. 1)	1.872	1.9985	1.9985	1.9985	1.9985	1.9985	1.9985
Journal Diameter, (No. 2)	1.7461	1.9985	1.9985	1.9985	1.9985	1.9985	1.9985
Journal Diameter, (No. 3)	1.247	1.9985	1.9985	1.9985	1.9985	1.9985	1.9985
Journal Diameter, (No. 4)	None	1.2485	1.2485	1.2485	1.2485	1.2485	1.2485
Piston Pin Diameter	.8592	.99965	.99965	.99965	.99965	.99965	.99965
Intake Valve Stem Diameter	.341	.3095	.30975	.3095	.30975	.3095	.30975
Exhaust Valve Stem Diameter	.3381	.3095	.30975	.3095	.30975	.3095	.30975
Top Compression Ring Width	1/8	1/8	1/8	1/8	1/8	1/8	1/8
Other Compression Ring Width	1/8	1/8	1/8	1/8	1/8	1/8	1/8
Oil Ring Width	1/4	3/16	3/16	3/16	3/16	3/16	3/16
Main Bearings, Diameter Clearance	0.2-2.4	1.5-3.5	2.3-4.5	1.5-3.5	2.3-4.5	1.5-2.5	2.3-4.5
Rod Bearings, Diameter Clearance	0.2-2.2	1.5-3.5	1.5-3.5	1.5-3.5	1.5-3.5	1.5-3.5	1.5-3.5
Piston Skirt Clearance, Aluminum	3		4-5		4-5		3.7-6.5
Piston Skirt Clearance, Iron		2.6-4.1		2.6-4.1		1.9-4.7	
Crankshaft End Play	4-6	2-7	2-6	2-7	2-6	2-7	2-6
Camshaft End Play	5-9	2-8	3-8	2-8	3-8	2-8	3-8
Camshaft Bearing Clearance	2-4	2-4.6	2-4.6	2-4.6	2-4.6	2-4.6	2-4.6
Crankcase Oil—U. S. Qts.	4	5	5	6	6	6	6
Crankcase Oil—Imp. Qts.	3.3	4	4	5	5	5	5
Cooling System—U. S. Gals.	3.12	3.5	3.5	4.75	4.75	4.75	4.75
Cooling System—Imp. Gals.	2.5	3	3	4	4	4	4
Transmission & Differential—U. S. Qts.	8	†20	†20	#40	#40	#40	#40
Transmission & Differential—Imp. Qts.	6.4	†17	†17	#32	#32	#32	#32
Final Drive Each—U. S. Pints	3						
Final Drive Each—Imp. Pints	2.4						
Pulley Housing—U. S. Qts.	***	2	2				
Pulley Housing—Imp. Qts.	***	1.75	1.75				

*Continental F 140 is used on current production tractors; early models used Continental F 124 which had a 3 inch bore, 4⅜ inch stroke and 124 cubic inch displacement.   ***Combined B.P. & P.T.O. housing holds 1.6 Imp. Qts. or 2 U. S. Qts.   #When equipped with P.T.O., add 10 Imp. pints or 12.8 U. S. pints.   †When equipped with live P.T.O. add 8.5 Imp. pints or 10.2 U. S. pints.

FRONT SYSTEM-TRICYCLE TYPE

SINGLE WHEEL FORK OR DUAL WHEEL AXLE AND SPINDLE

Model 20

1. Dual front wheels are mounted on a horizontal axle which is welded to the pedestal as shown in Fig. CS1. The procedure for removing the pedestal and axle assembly is evident after an examination of the unit. When installing the wheel hubs, tighten bearing adjusting nut (11) enough to give the taper roller bearings a very slight rotational drag.

Models 30-40 (Single Wheel)

2. The single wheel fork is flange bolted to the steering spindle as shown in Fig. CS3 and the removal procedure is evident. The taper roller bearings (8) should be adjusted to provide a very slight rotational drag.

Models 30-40-50 (Dual Wheels)

3. Dual front wheels are mounted on a horizontal axle which is welded to the vertical steering shaft or spindle as shown in Fig. CS5. To remove the axle and spindle unit from tractor, first remove grills and front hood and on early model 30 tractors, remove the front mounted hydraulic tank and support. Support front of tractor in a hoist and remove wheel and hub units. Remove cover from steering gear housing, remove nut or snap ring from upper end of spindle and bump spindle down. Raise front of tractor and withdraw spindle and axle unit from below. The bushing (11) can be renewed at this time. After installation, ream the bushing to an inside diameter of 1.750-1.751. This will provide a clearance of 0.001-0.003 for the 1.748-1.749 diameter shaft.

Reassemble the unit by reversing the disassembly procedure and adjust the taper roller axle bearings to provide a slight amount of rotational drag.

Note: When installing the steering shaft or spindle, make certain that front wheels are straight ahead when steering gear is in the mid-position.

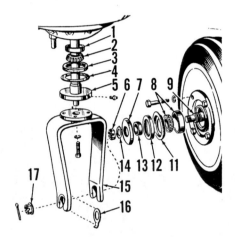

Fig. CS3—Model 40 single wheel fork and related parts. The model 30 single front wheel arrangement is similar, except washer (4) is not used.

1. Bearing cup	8. Bearing
2. Bearing cone	9. Axle
3. Felt washer	11. Oil seal ring
4. Washer (Model 40)	12. Oil seal
5. Steering post or shaft	13. Spacer
6. Nut	14. Lock washer
7. Dust cover	15. Wheel fork
	16. Tab washer

PEDESTAL

Models 30-40-50 (Dual Wheel)

4. **REMOVE AND REINSTALL.** To remove the pedestal on model 30 or the lower pedestal on models 40 and 50, first remove the horizontal axle and vertical spindle as outlined in paragraph 3. On models 40 and 50, unbolt upper pedestal from lower pedestal. On model 30, unbolt steering gear housing from pedestal. On all models, remove cap screws retaining lower pedestal to frame and remove pedestal.

Refer to paragraph 3 for information concerning the pedestal bushing and notes regarding the installation of the steering spindle.

FRONT SYSTEM-AXLE TYPE

STEERING KNUCKLES

Models 20-30-40-50

7. The procedure for removing the Lemoine type knuckles is evident after an examination of the unit and reference to Fig. CS7, CS9 or CS11. Spindle bushings (A) can be driven from the axle or axle extensions and new ones pressed in. Replacement bushings must be align reamed after installation to the inside diameter which follows:

Model 20 .... 1.241-1.243

Models 30, 40 & 50 (Adj. Axle)
.... 1.358-1.360

Models 30, 40 & 50 (Non-Adj. Axle) 1.3585-1.3595

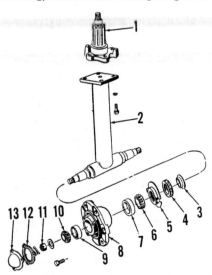

Fig. CS1 — Model 20 dual wheel tricycle pedestal and associated parts. Item (1) is the same as item (18—Fig. CS13).

1. Steering post	6 & 7. Bearing
2. Pedestal	8. Hub
3. Carrier for felt seal	9. Bearing cup
4. Felt seal	10. Bearing cone
5. Retainer for felt seal	11. Nut
	12. Gasket

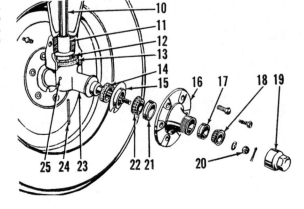

Fig. CS5—Model 30 dual wheel tricycle front system. Models 40 and 50 are similarly constructed. Items (10) and (23) are a welded unit.

10. Steering post or shaft
11. Bushing
12. Thrust washer
13. Felt washer
14. Washer
15. Seal retainer
16. Hub
17. Bearing cup
18. Bearing cone
19. Cap
20. Nut
21. Bearing cup
22. Bearing cone
23. Axle
24. Key
25. Pin

COCKSHUTT 20-30-40-50

Paragraphs 7-11

New spindle post diameter and the recommended bushing to spindle clearances are as follows:

Spindle post diameter:
Model 20 .... 1.239-1.240
Models 30, 40 & 50 (Adj. Axle)
.... 1.3565-1.3575
Models 30, 40 & 50
(Non-Adj. Axle) 1.3565-1.3575

Recommended bushing to spindle clearance:
Model 20 .... 0.001-0.004
Models 30, 40 & 50 (Adj. Axle)
.... 0.0005-0.0035
Models 30, 40 & 50
(Non-Adj. Axle) 0.001-0.003

When installing the wheel and hub units, tighten the adjusting nuts enough to give the taper roller bearings a slight rotational drag.

TIE-RODS
Models 20-30-40-50

8. The procedure for renewing the tie rods and/or tie rod ends is self-evident. When reassembling, adjust the length of each tie-rod an equal amount, to provide a recommended toe-in of 1/8 inch.

AXLE PIVOT PINS AND BUSHINGS
Model 20

9. To remove the axle pivot pin (20—Fig. CS7), raise front of tractor enough to remove weight from front wheels, remove lock plate (34) and bump pivot pin out of axle and pivot bracket. To renew bushings (22), disconnect tie-rods from center steering arm (pitman arm) and slide axle assembly far enough to one side to make the bushings accessible; then, drive bushings from the axle center member. The axle pivot pin bushings, which are interchangeable with the spindle post bushings (A), must be align reamed after installation to an inside diameter of 1.241-1.243. The 1.239-1.240 diameter pivot pin should have a recommended clearance of 0.001-0.004 in the bushings.

Models 30-40-50
(Non-Adjustable Axle)

10. To remove the axle pivot pin (15—Fig. CS9), first disconnect the radius rod anchor bracket (23) from the engine frame; then, disconnect both tie rods from the center steering arm (pitman arm). Support front of tractor enough to remove weight from front wheels and bump axle and radius rod assembly rearward and off the axle pivot pin. Using a punch or drift, drive the locking pin (16) from pedestal and bump the pivot pin (15) rearward and out of pedestal. The axle pivot pin bushing (17), which is interchangeable with the spindle post bushings (A), must be reamed after installation to an inside diameter of 1.355-1.357. The pivot pin should have a recommended clearance of 0.0015-0.0045 in the bushing.

Models 30-40-50
(Adjustable Axle)

11. To remove the axle pivot pins and bushings, disconnect tie rods from center steering arm (pitman arm) and support front of tractor enough to remove weight from front wheels. Remove cotter pin and nut from hinge pin (21—Fig. CS11) and remove the pin; then, bump axle assembly off the rear pivot pin (1). Rear pivot pin (1) and pivot bushings (11 & 13) can be renewed at this time. After installation, ream bushing (11) to an inside diameter of 1.999-2.000, and bushing

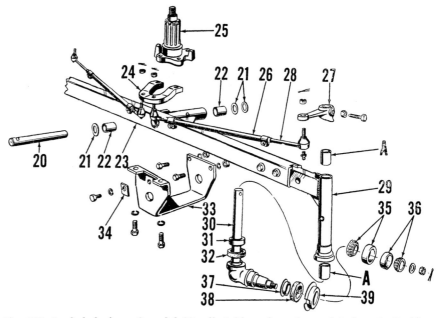

Fig. CS7—Exploded view of model 20 adjustable axle and associated parts. Bushings (A & 22) require final sizing after installation.

A. Spindle bushings (same as 22)
20. Axle pivot pin
21. Spacer washers
22. Pivot pin bushings
23. Axle center member
24. Steering (Pitman) arm
25. Steering post
26. Tie rod tube
27. Steering arm
28. Tie rod end
29. Outer axle member
30. Spindle and king pin or knuckle
31. Thrust bearing
32. Felt seal
33. Axle pivot bracket
34. Locking plate
35. Inner bearing
36. Outer bearing
37. Carrier for felt seal
38. Felt seal
39. Retainer for felt seal

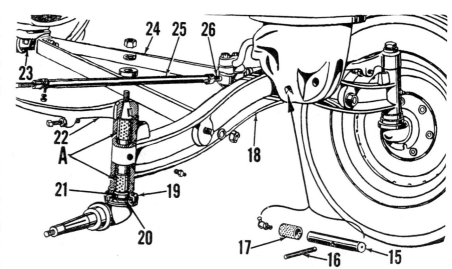

Fig. CS9—Model 30 non-adjustable front axle. Models 40 and 50 are similar. Bushings (A & 17) require final sizing after installation.

A. Spindle bushings (same as 17)
15. Axle pivot pin
16. Locking pin
17. Pivot pin bushing (same as A)
18. Axle
19. Bearing cover
20. Felt washer
21. Thrust bearing
22. Steering arm
23. Radius rod anchor bracket
24. Radius Rod
25. Tie rod tube
26. Tie rod end

CS-23

Paragraphs 11-20

COCKSHUTT 20-30-40-50

(13) to an inside diameter of 1.4822-1.4837. The 1.480-1.481 diameter rear pivot pin should have a clearance of 0.0012-0.0037 in the rear bushing (13). The 1.995-1.996 diameter spacer (12) should have a clearance of 0.003-0.005 in front bushing (11).

RADIUS ROD
Models 30-40-50
(Non-Adjustable Axle)

12. To remove the radius rod and anchor bracket assembly, remove the three cap screws securing the anchor bracket (23—Fig. CS9) to the engine frame. Remove the bolts retaining the radius rod to the axle and withdraw the radius rod and anchor bracket assembly. The anchor bracket can be removed from the radius rod if renewal is required. To renew only the anchor bracket, it is not necessary to disconnect radius rod from axle.

PIVOT BRACKET OR LOWER PEDESTAL
Model 20

13. **REMOVE AND REINSTALL.** To remove the axle pivot bracket (33—Fig. CS7), first raise front of tractor enough to remove weight from front wheels, remove lock plate (34) and bump pivot pin out of axle and pivot bracket. Unbolt pivot bracket from the nose casting and the engine front mounting bracket.

Note: If difficulty is encountered when reinstalling the bracket, support engine under oil pan and loosen all bolts retaining the engine front mounting bracket. Retighten after all bolts are started loosely.

Models 30-40-50

14. **R&R AND OVERHAUL.** To remove the axle pivot bracket or lower pedestal from tractor, first disconnect both tie rods from the center steering arm (pitman arm). On non-adjustable type axles, disconnect the radius rod anchor bracket (23—Fig. CS9) from the engine frame, raise front of tractor enough to remove weight from front wheels and bump the axle and radius rod assembly rearward and off the axle pivot pin. On the adjustable axle versions, raise front of tractor, remove hinge pin (21—Fig. CS11) and bump axle assembly off the rear pivot pin. On models 40 and 50, disconnect the center steering arm from the vertical steering shaft. On all models, remove grilles. On model 30, disconnect steering gear housing from pedestal. On models 40 and 50, disconnect upper pedestal from lower pedestal. On all models, unbolt the pivot bracket or pedestal from engine frame and withdraw the unit. The need and procedure for further disassembly is evident after an examination of the unit.

14A. The pivot bracket or lower pedestal on models 40 and 50 is fitted with a bushing (22—Fig. CS11) to support the lower end of the vertical steering shaft. The bushing is supplied, already installed in a new pivot bracket or lower pedestal; or, the bushing is available as an individual repair item. After installation, ream the bushing to an inside diameter of 1.4822-1.4847. The 1.4785-1.4797 diameter vertical steering shaft should have a clearance of 0.0025-0.0062 in the bushing.

STEERING GEAR
Model 20

As shown in Fig. CS12, the steering gear housing is an integral part of the nose casting to which is bolted the tractor frame side rails. For the purposes of this manual, the steering gear unit will include all of the parts exploded from the nose casting in Fig. CS15.

20. **ADJUSTMENT.** Three adjustments are provided on the steering gear unit: (1) The wormshaft bearing adjustment; (2) The backlash between the worm and worm wheel and (3) The vertical steering shaft (steering post) bearing adjustment.

Note: Before attempting to adjust the steering gear, raise front of tractor to remove any unnecessary load from the gear unit. Remove grilles and if more working room is desired, remove front hood.

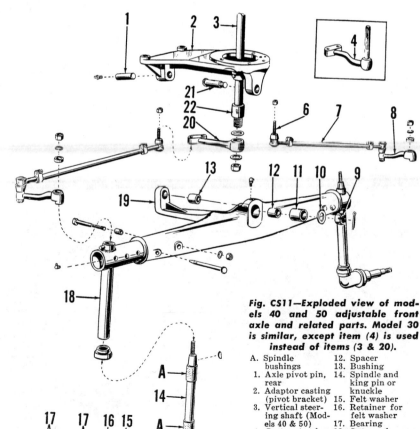

Fig. CS11—Exploded view of models 40 and 50 adjustable front axle and related parts. Model 30 is similar, except item (4) is used instead of items (3 & 20).

A. Spindle bushings
1. Axle pivot pin, rear
2. Adaptor casting (pivot bracket)
3. Vertical steering shaft (Models 40 & 50)
4. Center steering arm & shaft (Model 30)
6. Tie rod end
7. Tie rod tube
8. Steering arm
9. Nut
10. Washer
11. Bushing
12. Spacer
13. Bushing
14. Spindle and king pin or knuckle
15. Felt washer
16. Retainer for felt washer
17. Bearing
18. Outer axle member
19. Axle center member
20. Steering (Pitman) arm (Models 40 & 50)
21. Hinge pin

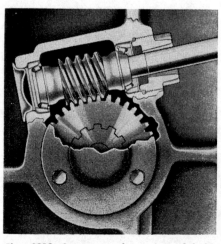

Fig. CS12—Cut-away view of model 20 nose casting, showing the steering worm and sector installation. See Fig. CS15 for exploded view.

COCKSHUTT 20-30-40-50

21. WORMSHAFT BEARING ADJUSTMENT. Remove cap screw and lock plate (1—Fig. CS13). Turn adjusting cap (2) either way as required to provide zero wormshaft end play, without causing binding in the taper roller bearings. Check the adjustment by turning the steering wheel through its full range of travel. Install the lock plate when adjustment is complete.

22. BACKLASH. The teeth of the worm wheel are tapered from top to bottom to provide a means of adjusting the backlash which should be equal to 2 inches free play at rim of steering wheel.

To adjust the backlash, proceed as follows: On adjustable axle versions, disconnect tie rods from the center steering arm (pitman arm) and on dual wheel tricycle versions, unbolt pedestal from the steering shaft (steering post). Remove the cap screws retaining the steering gear housing bottom cover (13—Fig. CS13) to the housing. Remove cover cap (7) from top of housing. Remove nut (9) and bump the steering shaft (steering post) and worm wheel assembly out through bottom of steering gear housing. Remove snap ring (6) and notice the number of shim washers (5) above and below the worm wheel. By relocating the various thickness of shims from above the worm wheel to below the worm wheel, or vice versa, the worm wheel can be moved up or down on the steering post and the larger portion of the tapered worm wheel teeth can be moved into or out of mesh with the worm; thus, changing the gear backlash. If backlash was excessive (more than 2 inches at rim of steering wheel), arrange the shims so as to bring the larger portion of the tapered worm wheel teeth further into mesh with the worm and reinstall the snap ring (6). It is important to remember, however, that one of the thicker shims **must** be next to the snap ring.

Note: It may be necessary to repeat this operation two or three times in order to obtain the desired backlash of 2 inches when measured at rim of steering wheel.

22A. Check for binding and/or backlash through entire range of steering wheel travel. If the gear unit binds or has an excessive amount of backlash in any position and the aforementioned adjustment will not correct the fault, it will be necessary to renew the worm and worm wheel or, re-position the worm wheel so as to bring unworn teeth into mesh. To re-position the worm wheel, proceed as follows: On adjustable axle versions, disconnect tie rods from the center steering arm (pitman arm) and on dual wheel tricycle versions, unbolt pedestal from the steering shaft (steering post). Remove the cap screws retaining the steering gear housing bottom cover to the housing. Remove cover cap (7—Fig. CS13) from top of housing. Remove nut (9) and bump worm wheel and shaft assembly out through bottom of housing. Turn the entire worm wheel and shaft assembly 180 degrees, reinstall the assembly and recheck the backlash as in paragraph 22.

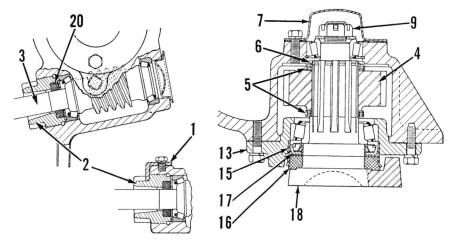

Fig. CS13—Sectional view of model 20 steering gear. The vertical shaft bearings are adjusted with nut (9). Worm shaft bearings are adjusted with cap (2). See Fig. CS15 for legend.

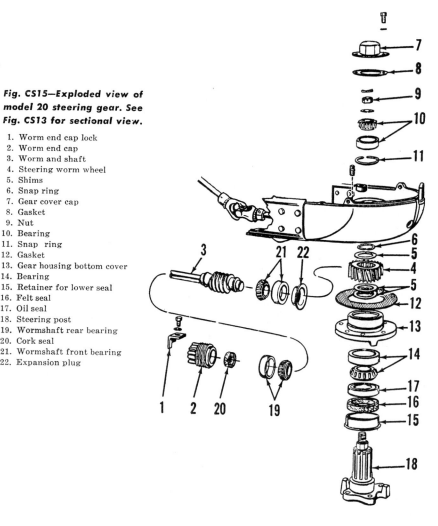

Fig. CS15—Exploded view of model 20 steering gear. See Fig. CS13 for sectional view.

1. Worm end cap lock
2. Worm end cap
3. Worm and shaft
4. Steering worm wheel
5. Shims
6. Snap ring
7. Gear cover cap
8. Gasket
9. Nut
10. Bearing
11. Snap ring
12. Gasket
13. Gear housing bottom cover
14. Bearing
15. Retainer for lower seal
16. Felt seal
17. Oil seal
18. Steering post
19. Wormshaft rear bearing
20. Cork seal
21. Wormshaft front bearing
22. Expansion plug

23. **VERTICAL SHAFT (STEERING POST) ADJUSTMENT.** Remove the steering gear housing top cover cap (7—Fig. CS13) and extract cotter pin from nut (9). Tighten the nut until all end play is removed from the vertical steering shaft and a very slight drag is obtained when turning the gear from one extreme position to the other. Tighten the nut one addition castellation and install the cotter pin.

24. **OVERHAUL.** The steering gear unit can be overhauled without removing the nose casting from the tractor, by proceeding as follows: Support front of tractor and on adjustable axle versions, disconnect center steering arm (pitman arm) from shaft (post) (18—Fig. CS15). On dual wheel tricycle versions, unbolt the pedestal from the steering shaft (18). Remove the cap screws retaining the gear housing bottom cover (13) to the housing. Remove the gear housing top cover cap (7), nut (9) and bump worm wheel and shaft assembly down and out of the housing. The removed unit can be disassembled by removing snap ring (6).

Using a punch, remove the lock pin retaining the steering shaft universal joint to the worm shaft. Grasp the steering wheel and pull the steering shaft and universal joint assembly rearward and out of way. Remove cap screw and lock plate (1). Unscrew adjusting cap (2) and withdraw the worm and shaft assembly. The need and procedure for further disassembly is evident after an examination of the unit.

When reassembling the steering gear, adjust the unit as outlined in paragraphs 20, 21, 22 and 23.

Model 30

Model 30 tractors are equipped with a Ross multiple stud type steering gear unit. The unit as used on the dual wheel tricycle, adjustable axle and non-adjustable axle versions is shown in Fig. CS17. The unit as used on the single wheel version is shown exploded in Fig. CS19.

25. **ADJUSTMENT.** The unit shown in Fig. CS17 is provided with two adjustments: (1) The worm (cam) shaft bearing adjustment and (2) The backlash between the worm (cam) and the lever studs. The unit shown in Fig. CS19, is provided with three adjustments: (1) The worm (cam) shaft bearing adjustment; (2) The backlash between the worm (cam) and the lever studs and (3) The steering spindle bearing adjustment.

Note: Complete adjustment of the steering gear unit cannot be accomplished without first removing the gear unit from the tractor as outlined in paragraph 30.

26. **WORM (CAM) SHAFT BEARING ADJUSTMENT.** Refer to Figs. CS17 or 19. To adjust the worm (cam) shaft bearings on all versions of the model 30, remove the gear unit as outlined in paragraph 30 and proceed as follows: On models so equipped, loosen seal clamp (2) and slide seal assembly rearward. Loosen the cap screws retaining the adjusting pad (5) to the lower side of the gear housing. Remove the cap screws retaining the gear housing rear cover to the housing and slide the cover rearward. Add or remove shims (4), which are available in thicknesses of 0.002, 0.003 and 0.010, until the worm (cam) shaft has a barely perceptible drag.

27. **BACKLASH.** To adjust the backlash between the worm (cam) and the steering lever studs on all versions of the model 30, proceed as follows: Remove adjusting pad (5) from underside of gear housing and vary the number of shims (6), which are available in thicknesses of 0.003, 0.007 and 0.010, until a slight drag is felt when turning the gear unit from one extreme position to the other.

28. **STEERING SPINDLE BEARING ADJUSTMENT.** When adjusting the complete steering gear unit on the single wheel tricycle versions, the steering spindle bearings should be adjusted before the steering gear unit is installed on the tractor; if, however, this is the only adjustment to be made, the work can be performed satisfactorily without removing the gear unit from the tractor by proceeding as follows:

Support front of tractor to remove load from steering gear, remove the gear housing top cover (7—Fig. CS19) and extract cotter pin from nut (9). Tighten nut (9) until all end play is removed from the steering spindle and a very slight drag is obtained when turning the gear from one extreme position to the other. Tighten the nut one additional castellation and install the cotter pin.

29. **OVERHAUL.** Complete overhaul of the steering gear unit cannot be accomplished without removing the gear unit from the tractor.

30. **REMOVE AND REINSTALL.** To remove the steering gear unit from the tractor, first remove the grilles, front hood and on early models, remove the front mounted hydraulic tank and support. Remove the baffle which is located under the radiator. On the single wheel tricycle version, support front of tractor and remove the wheel fork assembly. On the adjustable and non-adjustable axle versions, disconnect tie rods from the center steering (pitman) arm. On the dual wheel tricycle version, remove the gear housing top cover, remove nut or snap ring from upper end of the verticle spindle and bump the spindle and wheels assembly down and out of the lower pedestal. On all versions, remove the lock pin retaining the steering shaft front universal joint to the worm shaft. Grasp the steering wheel and pull and bump the steering shaft rearward until the universal joint is free from the worm shaft.

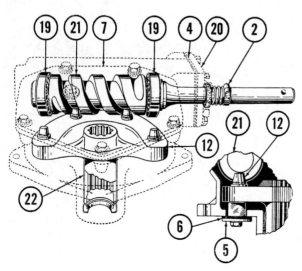

Fig. CS17 — Model 30 steering gear, as used on the dual wheel tricycle, adjustable axle and non-adjustable axle versions. Refer to Fig. CS19 for the single wheel tricycle version.

2. Seal clamp (some models)
4. Shims
5. Adjusting pad
6. Shims
7. Gear housing cover
12. Steering lever
19. Bearing
20. Cap
21. Worm
22. Bushing

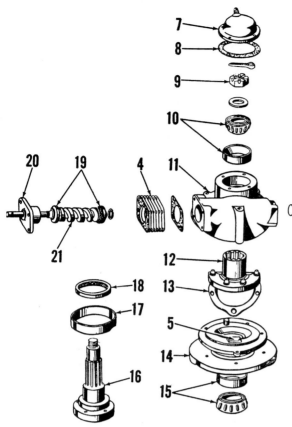

Fig. CS19—Exploded view of model 30 steering gear as used on the single wheel tricycle version. See Fig. CS17 for the gear unit as used on other versions.

4. Shims
5. Adjusting pad
7. Gear housing cover
8. Gasket
9. Nut
10. Bearing
11. Steering gear housing
12. Steering lever
13. Gasket
14. Gear housing bottom cover
15. Bearing
16. Steering spindle
17. Dust shield
18. Oil seal
19. Bearings
20. Cap
21. Worm

Unbolt the gear housing from the adaptor casting (axle pivot bracket) or lower pedestal and withdraw the gear housing from the tractor.

Reinstall the unit by reversing the removal procedure and make certain that middle studs of the steering lever are in mesh with the worm as shown in Fig. CS20, when front wheel (or wheels) are straight ahead.

31. DISASSEMBLE AND REASSEMBLE. The steering gear unit can be disassembled after first removing the gear unit from the tractor as outlined in the preceding paragraph. Remove the gear housing rear cover (20—Figs. CS 17 or 19) and turn the worm shaft out of the housing. On the dual wheel tricycle versions, the steering lever (12—Fig. CS 17) can be removed at this time. On the single wheel tricycle version, remove the gear housing top cover (7—Fig. CS 19), remove nut (9) and bump the spindle (16) down and out of the housing. Unbolt the gear housing lower cover (14) from the housing, remove the cover and withdraw the lever (12). On the adjustable and non-adjustable axle versions, remove the gear housing top cover, remove nut or snap ring from the vertical spindle (spline), bump the spindle and center steering (pitman) arm unit out of the steering lever and withdraw the lever.

Fig. CS20—Model 30 steering gear as used on the dual wheel tricycle, adjustable axle and non-adjustable versions. The middle studs of the steering lever should be meshed with the steering worm when front wheels are straight ahead.

Fig. CS21—Models 40 and 50 steering gear. See Fig. CS23 for an exploded view.

1. Gear housing cover
2. Spring
3. Oil seal cup
5. Wormshaft rear bearing
6. Nut
7. Tab washer
8. Sector
9. Worm
10. Wormshaft front bearing
11. Shims
12. Eccentric bushing
13. "O" ring
14. Vertical steering shaft
15. Adjusting cap
16. Shims
17. Set screw
18. Cotter pin

The need and procedure for further disassembly is evident after an examination of the unit. The steering gear housing on all models except the single wheel tricycle version is fitted with a bushing (22—Fig. CS 17) which carries the lower portion of the steering lever. This bushing should be reamed after installation, if necessary, to provide a free fit for the 1.362-1.363 diameter lever shaft.

Reassemble the gear unit by reversing the disassembly procedure, make certain that middle studs of the steering lever are in mesh with the worm as shown in Fig. CS 20 when front wheel (or wheels) are pointing straight ahead and adjust the unit as outlined in paragraphs 25, 26, 27 and 28.

Models 40-50

All versions of the 40 and 50 tractors are equipped with the worm and sector type steering gear unit as shown in Figs. CS21 and CS23.

32. ADJUSTMENT. Three adjustments are provided on the steering gear unit: (1) The worm shaft bearing adjustment; (2) The vertical steering shaft (steering post) end play and (3) The backlash between the worm and sector.

Note: Before attempting to adjust the steering gear, raise front of tractor so as to remove any unnecessary load from the unit and remove the grilles and front hood.

33. WORMSHAFT BEARING ADJUSTMENT. To adjust the wormshaft bearings, remove adjusting cap (15—Fig. CS21 or 23) and vary the number of shims (16), which are available in thicknesses of 0.002, 0.003 and 0.010, until the worm shaft has a

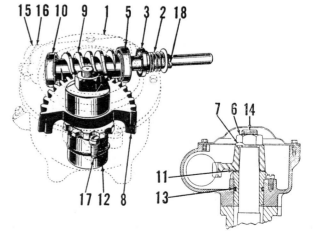

Paragraphs 33-41   COCKSHUTT 20-30-40-50

slight amount of drag. Note: A maximum pre-load on the bearings equivalent to a two pound pull exerted at the rim of the steering wheel is permissible.

34. VERTICAL SHAFT END PLAY. To adjust the vertical steering shaft (steering post) end play, first remove the gear housing top cover (1—Fig. CS 21 or 23); then, remove nut and tab washer (7). Remove sector (8) and vary the number of shims (11), which are available in thicknesses of 0.010 and 0.020, to remove any excessive end play. The desired adjustment is zero end play without binding. Since only 0.010 and 0.020 shims are available as a stock item, it may not be possible to remove all vertical shaft end play without causing binding. In which case, a slight amount of end play is permissible.

When adjustment is complete, make certain that nut (6) is tightened securely and that tab washer (7) is turned up against the nut.

35. BACKLASH. To adjust the backlash between the worm and sector, first place the front wheel (or wheels) in the straight ahead position and remove the gear housing top cover. Remove set screw (17—Fig. CS21) and using a punch and hammer, turn eccentric bushing (12) either way as required to reduce the backlash between the worm and sector to one inch when measured at rim of steering wheel. When adjustment is complete, install set screw (17).

36. OVERHAUL. Overhaul of the steering gear unit can be accomplished without removing the unit from the tractor.

36A. Remove grilles, steering wheel lock pin and steering wheel. Remove front and center hood as a unit. Remove the gear housing top cover (1—Fig. CS23) and adjusting cap (15). Remove nut (6) and withdraw sector (8). Extract cotter pin (18) and withdraw steering shaft from front of housing as shown in Fig. CS24.

36B. The need and procedure for further disassembly is evident after an examination of the unit. The 1.4785-1.4797 diameter vertical steering shaft (steering post) should have a clearance of 0.0033-0.0055 in the eccentric bushing (12).

When reassembling, renew "O" ring (13) on eccentric bushing. Adjust the steering gear unit as outlined in paragraphs 32, 33, 34 and 35.

37. R&R HOUSING. To remove the steering gear housing and associated parts, first perform the work outlined in paragraph 36A. Then, unbolt the gear housing from the upper pedestal and remove the housing.

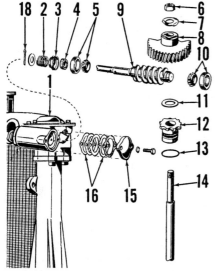

Fig. CS23—Exploded view of models 40 and 50 steering gear. Refer to Fig. CS24 for removal of worm (9). See Fig. CS21 for legend.

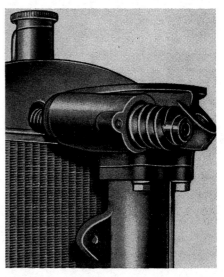

Fig. CS24—Models 40 and 50 steering wormshaft can be withdrawn from the housing after removing the steering wheel and cotter pin (18—Fig. CS23).

ENGINE AND COMPONENTS

R&R ENGINE WITH CLUTCH

Model 20

40. To remove the engine and clutch as a unit, first drain cooling system and if engine is to be disassembled, drain oil pan; then, remove grilles, remove front and center hood as a unit and proceed as follows: Remove battery cover, disconnect battery cables and remove the battery. Remove deck plate (sheet metal covering over clutch shaft) from tractor. Remove clevis pin from rear end of clutch release rod and remove the release rod. Disconnect the clutch shaft universal joint and slide clutch shaft rearward. Unbolt clutch housing cover and remove cover and clutch shaft as a unit. Disconnect cable from starting motor, disconnect wires from voltage regulator and coil and pull wiring harness rearward and out of way. Shut off fuel, disconnect fuel line (or lines), unbolt fuel tank support from clutch housing and remove main fuel tank and support from tractor. Disconnect the steering shaft universal joint and pull the steering shaft rearward enough to clear the breather tube which is welded to the valve chamber cover. Disconnect the governor control rod at governor, choke control rod at carburetor and on distillate models, disconnect radiator shutter control wire at shutters and pull wire rearward and out of way. Disconnect radiator hoses and heat indicator bulb. Drain hydraulic system, disconnect hydraulic lines from pump and remove clamp securing hydraulic lines to clutch housing.

Support engine in a hoist, unbolt the tractor frame left channel and remove channel. Remove the remaining cap screws and bolts securing engine to frame and move engine rearward and away from tractor.

The extent of the work done after engine removal will dictate, to a certain extent, the installation procedure. In general, however, the engine can be installed by reversing the removal procedure.

Model 30

41. To remove the engine and clutch as a unit, first drain cooling system and if engine is to be disassembled, drain oil pan; then, remove grilles, remove front and center hood as a unit and proceed as follows: On early models, remove the front mounted hydraulic tank and support. Disconnect radiator hoses from engine, remove the sheet metal baffle plate which is located under radiator and on distillate models, disconnect the radiator shutter control wire at shutters. Remove radiator and on models so equipped, disconnect hour meter wires from oil pressure switch. On non-Diesel models, disconnect lines from hydraulic pump, remove battery box cover, disconnect battery cables from battery and remove battery and battery box. On all models, disconnect wire from voltage regulator, cable from starting motor and heat indicator bulb from the water outlet casting on top of cylinder head. On gasoline and distillate models, shut off fuel and disconnect fuel line and choke rod from carburetor. On all non-Diesel models, disconnect rod from starting motor, disconnect governor control rod at governor, disconnect distributor wire from coil and remove the carburetor air intake tube. On all models, disconnect oil gage line at junction block on left side of cylinder head, unbolt tail light wire clips and disconnect tail light wire at tail light. On L.P. gas models, disconnect and remove water pipes, vapor hose and balance line from converter. On Diesel models, remove the manifold air intake tube, disconnect heater cable from manifold and disconnect speed control rod (or rods) and the fuel supply and return lines. On models so equipped, remove tool box from right side of engine. Using a punch and hammer, remove the pin retaining the rear universal joint to the steering shaft and bump the universal joint free from the shaft. Disconnect fuel tank brackets and lift fuel tank, steering wheel and shaft, instrument panel, etc., as a unit, from tractor as shown in Fig. CS31. Disconnect engine front mount from frame and be careful not to mix or lose shims which may be installed between the mount and the frame. Remove clutch housing cover and starting motor. Remove the engine rear mounting bolts and using a drift, bump both of the engine locating dowels up and out of the engine frame. Attach hoist to engine in a suitable manner and lift engine from tractor.

The procedure for installing the engine will depend to some extent on the work which was done after the engine was removed; but in general, the engine can be installed by reversing the removal procedure.

Models 40-50

42. To remove the engine and clutch as a unit, first drain cooling system and if engine is to be disassembled, drain oil pan; then, remove grilles, remove front and center hood as a unit and proceed as follows: Remove steering wheel. Remove adjusting cap from front of steering gear housing and withdraw cotter pin from steering shaft. Turn the steering worm shaft out of mesh with the sector and

Fig. CS31 — Model 30 instrument panel, fuel tank, etc., removed from tractor in preparation for R & R of engine.

Fig. CS30 — Model 20 engine and clutch removed from tractor. The clutch housing has been removed in preparation for removing flywheel. The clutch, however, can be removed without removing the engine or the clutch housing.

remove worm shaft from tractor. Disconnect radiator hoses and remove the radiator and the fan. On models so equipped, disconnect hour meter wires from oil pressure switch. Remove oil pan cap screw retaining the wire clip on right side of engine. Disconnect cable from starter, wire from voltage regulator and heat indicator bulb from the water outlet casting on top of cylinder head. Disconnect oil gage line at junction block on left side of cylinder head and on non-Diesel models, disconnect wire from ignition coil and disconnect governor control rod. On gasoline models, shut off fuel, disconnect choke rod and fuel line from carburetor and remove the carburetor air inlet tube. On L.P. gas models, disconnect and remove water pipes and vapor hose from converter. On Diesel models, remove the manifold air intake tube, disconnect heater cable from manifold and disconnect speed control rod (or rods) and the fuel supply and return lines. Disconnect bracket from the flywheel housing plate and block up under fuel tank assembly. Remove clutch housing cover. Disconnect engine front mount from frame and be careful not to mix or lose shims which may be installed between the mount and the frame. Remove the engine rear mounting bolts and using a drift, bump both of the engine locating dowels up and out of the engine frame. Attach hoist to engine in a suitable manner and lift engine from tractor.

The extent to which the engine is disassembled subsequent to its removal will, to some extent, govern the installation procedure. In general, however, the engine can be installed by reversing the removal procedure.

CYLINDER HEAD
Model 20

43. To remove the cylinder head, first drain cooling system and remove center hood. Note: If more working room is desired, first remove grilles, then remove front and center hood as a unit. Remove the ignition distributor and the distributor drive shaft. Remove air cleaner and the carburetor air intake tube. Disconnect oil filter base from cylinder head and remove the head.

When installing the cylinder head, tighten the nuts from the center outward and to a torque of 70-75 Ft.-Lbs. Check the ignition timing as outlined in paragraph 201.

Models 30-40-50 (Non-Diesel)

44. To remove the cylinder head, first drain cooling system and remove center hood. Note: If more working room is desired, first remove grilles, then remove front and center hood as a unit. Disconnect the oil pressure lines from junction block on left side of cylinder head and disconnect heat indicator bulb from water outlet casting on top of head. Disconnect carburetor from manifold and on L.P. gas models, remove water lines which connect to converter. On distillate models, remove the auxiliary fuel tank. On all models, disconnect water pump from head. Disconnect by-pass hose from thermostat housing, remove cap screw retaining the water inlet pipe to cylinder head and lay by-pass line and inlet pipe assembly forward and out of the way. Remove thermostat housing from water outlet casting and water outlet casting from head. Remove valve cover, rocker arms assembly and push rods. Remove the cylinder head retaining stud nuts and lift cylinder head from tractor.

When installing the cylinder head, tighten the stud nuts progressively, from center outward and to a torque value of 95-100 Ft.-Lbs. Adjust the valve tappet gap to 0.012 hot.

Models 30-40-50 (Diesel)

45. To remove the cylinder head, first drain cooling system and remove center hood. Note: If more working room is desired, first remove grilles, then remove front and center hood as a unit. Disconnect the oil pressure lines from junction block on left side of head and on models 40 and 50, remove the oil line which runs from the oil gallery to the junction block. Disconnect heat indicator bulb from water outlet casting on top of head and disconnect water pump from front of head. Disconnect by-pass hose from thermostat housing, remove cap screw retaining the water inlet pipe to cylinder head and lay by-pass line and inlet pipe assembly forward and out of the way. Remove thermostat housing from water outlet casting and water outlet casting from head. Disconnect heater cable from manifold and remove the manifold air inlet tube. Disconnect the Diesel fuel system supply and return lines, disconnect the second and third fuel filters mounting bracket and lay filters and bracket over and out of way. Disconnect nozzle lines from injection pump and on models 40 and 50, loosen number 6 nozzle and disconnect the injection line from the nozzle. Remove valve cover, rocker arms assembly and push rods. Remove the cylinder head retaining stud nuts and lift cylinder head from tractor.

When installing the cylinder head, tighten the stud nuts progressively, from center outward and to a torque value of 95-100 Ft.-Lbs. Adjust the valve tappet gap to 0.012 hot.

VALVES AND SEATS
Model 20

46. Intake and exhaust valves are not interchangeable and the intake valves seat directly in the cylinder block with a face and seat angle of 30 degrees. Exhaust valves seat on renewable ring type inserts with a face and seat angle of 45 degrees. Desired seat width is 1/16 inch for the intake, 5/64 inch for the exhaust. Valve seats can be narrowed, using 20 and 70 degree stones. Refer to Fig. CS34 for method of removing the renewable type valve seat inserts. Valves have a stem diameter of 0.3406-0.3414 for the intake, 0.3377-0.3385 for the exhaust.

Tappet adjustment is a three wrench job as shown in Fig. CS35. Adjust both the intake and exhaust tappet gap to 0.014 hot. If tappet gap is adjusted with engine cold, allow approximately 0.002 for expansion purposes and recheck gap when engine is hot.

Models 30-40-50

47. Intake and exhaust valves are not interchangeable and the intake valves on all models seat directly in cylinder head with a face and seat angle of 45 degrees. The exhaust valves in production non-Diesel engines seat directly in cylinder head with a face and seat angle of 45 degrees. The Diesel engine exhaust valves seat on renewable ring type inserts with a face and seat angle of 45 degrees. The method of removing the renewable type valve seat inserts is shown in Fig. CS34. Standard production Diesel engine exhaust valve seat inserts (Cockshutt part No. T-5438) can be used as a repair item on non-Diesel engines, providing suitable counter-boring equipment is available. Measure outside diameter of a new seat (at room temperature) and cut counter-bore in head 0.002 smaller than the measured O.D. of the seat insert; then, chill the seat insert in dry ice to facilitate installation. Desired valve seat width is 1/16-1/8 inch for both intake and exhaust. Seats can be narrowed, using 20 and 70 degree

Fig. CS34—Method of removing a valve seat insert on Diesel models 30, 40 and 50. The same technique can be used on other models equipped with seat inserts.

Fig. CS35—Model 20 tappet adjustment is a three wrench job as shown. The desired tappet gap is 0.014 hot.

stones. Intake and exhaust valves have a stem diameter of 0.309-0.310 for non-Diesel engines, 0.3095-0.3100 for Diesel engines.

Adjust the intake and exhaust tappet gap to 0.012 hot.

VALVE GUIDES AND SPRINGS
Model 20

48. Intake and exhaust valve guides (5—Fig. CS37) are interchangeable and can be driven from cylinder block if renewal is required. Press guides into cylinder block with smaller O.D. of guide down, until port end of guide is 1 15/32 inches below machined top surface of cylinder block. Maximum allowable clearance limits between the valve stems and the guides is 0.0008-0.0026 for the intake, 0.0037-0.0055 for the exhaust. Ream new guides after installation, to provide the recommended stem to guide clearance of 0.0015 for the intake, 0.0045 for the exhaust. Note: The same size reamer can be used for both the intake and exhaust valve guides, since the difference in clearance specifications is accounted for by the differences in the valve stem diameters. Refer to paragraph 46 for the valve stem diameter.

49. Intake and exhaust valve springs (6—Fig. CS37) are interchangeable. Renew any spring which is rusted, discolored or does not meet the load test specifications which follow:

Pounds pressure at 1 45/64 inches
.............................47-53
Pounds pressure at 1 27/64 inches
.............................96-104

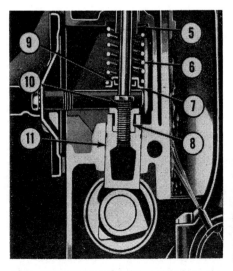

Fig. CS37—Sectional view showing the model 20 valve, tappet and camshaft installation.

5. Valve guide
6. Valve spring
7. Retainer for spring
8. Lock nut
9. Valve spring retainer lock
10. Adjusting screw
11. Tappet

Models 30-40-50

50. New intake and exhaust valve guides are interchangeable. Guides can be pressed or driven from cylinder head if renewal is required. Press guides into cylinder head with smaller O.D. of guide up, until the distance from the top of the guide to bottom of the valve spring counterbore in top of cylinder head is 13/16 inch for non-Diesels, 3/4 inch for Diesels. Maximum allowable valve stem to guide clearance is 0.006. Ream new guides after installation, to provide the recommended stem to guide clearance of 0.0015-0.0035 for non-Diesels, 0.0015-0.003 for Diesels. Refer to paragraph 47 for valve stem diameter.

51. Intake and exhaust valve springs are interchangeable in all models. Renew any spring which is rusted, discolored or does not meet the load test specifications which follow:

Spring free length......2 3/32 inches
Pounds pressure at 1 27/32 inches
.............................42-47
Pounds pressure at 1 13/32 inches
.............................122-131

VALVE TAPPETS
(CAM FOLLOWERS)
Model 20

52. The 0.999-0.9995 diameter barrel type tappets (11—Fig. CS37) ride directly in the unbushed cylinder block bores with a suggested clearance of 0.0005-0.0015.

Any one tappet can be renewed without disturbing cylinder head or camshaft by proceeding as follows: Remove the valve chamber cover, loosen the tappet adjusting screw lock nut and screw the tapped adjusting screw (10) completely down. Remove the valve spring retaining pin, hold valve up and remove the

Fig. CS39—Models 30, 40 and 50 rocker arms assembly is lubricated via the slotted stud (S).

valve spring. Note: A hooked wire inserted through the spark plug hole will facilitate holding the valve up. While holding the valve up, remove the tappet adjusting screw and lock nut from tappet and lift tappet out of its bore. Tappets are available in standard size only.

Note: If more than one tappet is to be removed, time will usually be saved by first removing cylinder head and valves.

When reassembling, adjust tappet gap to 0.014 hot.

Models 30-40-50

53. The 0.560-0.561 diameter mushroom type tappets operate directly in the unbushed cylinder block bores with a suggested clearance of 0.0005-0.0025. Maximum allowable clearance is 0.0035.

To remove the tappets, it is first necessary to remove the camshaft as outlined in paragraph 63. Tappets are available in standard size only.

When reassembling, adjust tappet gap to 0.012 hot.

ROCKER ARMS
Models 30-40-50

54. The hollow rocker arm shaft is drilled for lubrication to each rocker arm bushing. Lubricating oil to the drilled cylinder head passage and slotted oil stud (S—Fig. CS39) is supplied by an external oil line which is connected to the main oil gallery on left side of engine. If oil does not flow from the hole in the top of each rocker arm, check for foreign material in the external oil line or in the cylinder head passage.

The procedure for disassembling and reassembling the rocker arms from the shaft is evident. Check the rocker arm shaft and the bushing in each rocker arm for excessive wear. Maximum allowable clearance between the shaft and bushings is 0.005. When installing new bushings, make certain that oil hole in bushing is in register with oil hole in rocker arm and ream the bushings to provide a clearance of 0.001-0.002 for the 0.8405-0.841 diameter rocker arm shaft. When installing the rocker arm shaft, make certain that the oil metering holes in the shaft point toward push rods instead of valve stems.

Inspect the valve stem contact button in the end of each rocker arm for being mutilated or excessively loose. If either condition is found, renew the contact button. Extract the button retaining snap ring as shown in Fig. CS40 and remove the button and oil wick. Install new oil wick and button and test the button for a free fit in the rocker arm socket.

Note: If a new contact button has any binding tendency in the rocker arm socket, use a fine lapping compound and hand lap the mating surfaces.

VALVE TIMING

Models 20-30-40-50

55. Valves are properly timed when timing marks on camshaft gear and crankshaft gear are in register. On model 20, the double punch marked tooth space (4—Fig. CS41) on the camshaft gear should be meshed with the single punch marked tooth (3) on the crankshaft gear. On models 30, 40 and 50, the single punch marked tooth space on the camshaft gear should be meshed with the single punch marked tooth on the crankshaft gear, as shown at (X) in Fig. CS42.

Valve timing data are as follows:
Model 20

Intake valve opens.....T.D.C.
Intake valve closes.....35°A.B.D.C.
Exhaust valve opens....40°B.B.D.C.
Exhaust valve closes....T.D.C.

Models 30, 40 & 50 (Non-Diesel)

Intake valve opens....10° B.T.D.C.
Intake valve closes....40° A.B.D.C.
Exhaust valve opens...45° B.B.D.C.
Exhaust valve closes... 5° A.T.D.C.

Models 30, 40 & 50 (Diesel)

Intake valve opens....20° B.T.D.C.
Intake valve closes....38° A.B.D.C.
Exhaust valve opens...45° B.B.D.C.
Exhaust valve closes...13° A.T.D.C.

CRANKCASE FRONT COVER

Model 20

56. To remove the crankcase front cover (timing gear cover), first drain cooling system and remove grilles; then, remove front and center hood as a unit. Disconnect radiator hoses and remove radiator. On models so equipped, drain the hydraulic system and remove the hydraulic pump. Disconnect speed control rod and throttle rod and remove governor. Loosen fan belt and remove fan blades. Remove crankshaft pulley and Woodruff key. Remove the dust shield which is bolted to the engine front support. Support engine under oil pan and remove the large cap screw which secures the crankcase front cover to the engine front support. Remove the

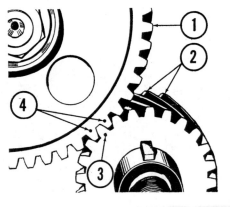

Fig. CS41—Model 20 valve timing marks. The double punch marked tooth space (4) on the camshaft gear (1) should be meshed with the single punch marked tooth (3) on the crankshaft gear (2).

Fig. CS40—Models 30, 40 and 50 rocker arms are fitted with renewable type valve stem contact buttons which can be removed after extracting the retaining snap rings as shown.

remaining cap screws and withdraw cover from front of engine. Notice which cap screws are equipped with copper washers so the washers can be installed in their original position.

The crankshaft front oil seal, which is retained in the crankcase front cover, can be renewed at this time. The spring loaded oil seal should be installed with lip of same facing inward toward timing gears.

When reassembling, leave the cover screws loose until after the crankshaft pulley is installed. This will facilitate centering the front oil seal with respect to the pulley. If difficulty is encountered when attempting to start the large cap screw which fastens the crankcase front cover to the engine front support, loosen the remaining support bolts, then start the cap screw.

Models 30-40-50

57. To remove the crankcase front cover (timing gear cover), first drain cooling system and remove grilles; then, remove front and center hood as a unit. On models 40 and 50, remove steering wheel. Remove the adjusting cap from front of steering gear housing, extract cotter pin from worm shaft and withdraw the worm shaft forward and out of steering gear housing as shown in Fig. CS42A. On model 30, remove the sheet metal baffle which is located under the radiator. On all models, disconnect radiator hoses and remove radiator. On non-Diesels, disconnect control rods and remove

Fig. CS42—Non-Diesel models 30, 40 and 50 timing gear train. The single punch marked tooth space on the camshaft gear should be meshed with the single punch marked tooth on the crankshaft gear as shown at (X). For the purposes of this illustration, the Diesel models are similar.

COCKSHUTT 20-30-40-50

governor. On all models, remove fan belt, crankshaft pulley and the pulley Woodruff key. Support engine under oil pan and remove the engine front mounting bracket (1—Fig. CS43), being careful not to mix or lose shims which may be installed between the mounting bracket and the frame. Remove the remaining cap screws and withdraw cover from front of engine.

The crankshaft front oil seal (2), which is retained in the crankcase front cover can be renewed at this time. The spring loaded oil seal should be installed with lip of same facing inward toward timing gears.

Reinstall the crankcase front cover by reversing the removal procedure and make certain that copper washers are installed on the bottom three cover retaining cap screws. On non-Diesels, refer to paragraph 191 for a caution concerning the proper installation of the governor shaft.

TIMING GEARS

Model 20

58. Either the camshaft gear and/or crankshaft gear can be removed without removing their respective shafts from the engine. To remove the gears, it is necessary to use a suitable puller after the crankcase front cover has been removed as outlined in paragraph 56.

Recommended timing gear backlash is 0.002-0.005. Renew the gears if the backlash is excessive. Timing gears are available in oversizes and undersizes of 0.001, 0.002, 0.003, 0.004, 0.005 and 0.006 to facilitate obtaining the desired backlash.

59. When installing the timing gears, the double punch marked tooth space on the camshaft gear should be meshed with the single punch marked tooth on the crankshaft gear as shown in Fig. CS41.

Note: When installing the camshaft gear, insert a heavy bar through the fuel pump hole opening in right side of crankcase to buck up the camshaft while the gear is being drifted on. This procedure will eliminate possible loosening of the soft plug which is located in cylinder block at rear of camshaft.

Models 30-40-50

60. Either the camshaft gear and/or crankshaft gear can be removed without removing their respective shafts from the engine. To remove the gears, it is necessary to use a suitable puller after the crankcase front cover has been removed as outlined in paragraph 57.

Before removing either of the gears, check the gear backlash which should be 0.001-0.003 for Diesel engines, 0.0005-0.0015 for non-Diesels. If the backlash exceeds 0.007, either one or both gears should be renewed. Before removing the camshaft gear and after removing the camshaft thrust plate retaining cap screws as shown in Fig. CS45, check the camshaft end play by inserting a feeler gage between the rear face of the camshaft thrust plate and the front face of the camshaft front bearing journal, as shown in Fig. CS46. The thickness of the feeler gage that can be inserted represents the camshaft end play which should be 0.003-0.008 for Diesel engines, 0.002-0.008 for non-Diesels. If the camshaft end play is excessive (more than 0.014), and a new thrust plate will not correct the condition, it will be necessary to renew the camshaft gear and recheck the end play or, if the same gear is to be used, it will be necessary to file the required amount from the rear face of the gear hub as

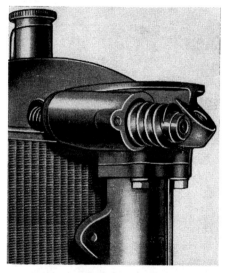

Fig. CS42A—Removing the steering worm shaft on models 40 and 50.

Fig. CS42B—Removing the crankshaft pulley on model 30. Models 40 and 50 are similar.

Fig. CS43—Models 40 and 50 timing gear cover and associated parts. Model 30 is similar.

1. Engine front support
2. Oil seal
3. Timing gear cover
4. Engine front plate

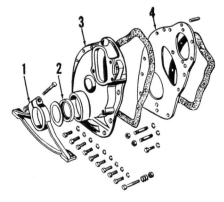

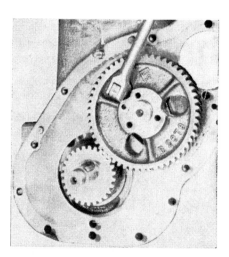

Fig. CS45—Removing the camshaft thrust plate retaining cap screws on models 30, 40 and 50.

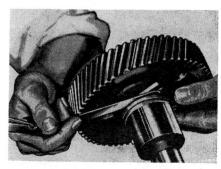

Fig. CS46—Checking the camshaft end play on models 30, 40 and 50. The amount of end play is equal to the thickness of the feeler gage that can be inserted between the shaft journal and the thrust plate.

shown in Fig. CS47. For example, if the measured end play is 0.015, file at least 0.007 from the gear hub so as to bring the end clearance within the suggested limits.

Timing gears are marked with a letter "S" if the gears are standard size; or, the gears are marked with a number within either an "O" or a letter "U". The letter "O" indicates an oversize gear and the letter "U" indicates an undersize gear. The enclosed number gives the deviation from a standard size gear in thousandths of an inch. When installing new gears, always use the same size gears as were removed and check the backlash to make certain that the value is within the clearance limits. Due allowance should be made, however, for wear on the other gear if only one gear is renewed.

Usually, new gears are punched with timing marks before they leave the factory. If, however, a new gear is not so marked, it is important to transfer the marks from the old gear to the new. To do so, proceed as follows: Place the old gear on top of the new one and using a piece of keystock, align the key ways of both gears. Place a straight edge against the gear teeth and in line with the timing mark on the old gear. Locate and punch mark the appropriate tooth (or tooth space) on the new gear. Refer to Fig. CS48.

To facilitate installation of a new crankshaft gear, boil the gear in oil for a period of 15 minutes prior to installation. When installing the camshaft gear, mesh the single punch marked camshaft gear tooth space with the single punch marked crankshaft gear tooth as shown in Fig. CS49.

Note: When installing the camshaft gear, insert a heavy bar through the fuel pump hole opening in left side of crankcase to buck up the camshaft while the gear is being drifted on. This procedure will eliminate possible loosening of the soft plug which is located in cylinder block at rear of camshaft.

CAMSHAFT AND BEARINGS
Model 20

The camshaft is supported in three steel-backed, babbitt-lined bearings. The front and rear shaft journals have a normal operating clearance of 0.002-0.004 in the bearings. The center journal has a normal operating clearance of 0.003-0.0045. If the journal clearance is excessive, the bearings and/or shaft should be renewed. To renew the bearings, follow the procedure outlined in paragraph 62. To renew the camshaft, refer to the following paragraph.

61. CAMSHAFT. To remove the camshaft, first remove the crankcase front cover as outlined in paragraph 56; then, proceed as follows: Remove the ignition distributor, distributor drive shaft and cylinder head. Disconnect choke rod, disconnect fuel line and remove carburetor. Remove the valve chamber cover, oil pan and oil pump. Lift the valves and hold in the raised position with wedges in a manner similar to that shown in Fig.

Fig. CS49—Non-Diesel models 30, 40 and 50 timing gear train. The single punch marked tooth space on the camshaft gear should be meshed with the single punch marked tooth on the crankshaft gear as shown at (X). For purposes of this illustration, the Diesel models are similar.

Fig. CS47 — Excessive camshaft end play on models 30, 40 and 50 can be corrected by filing the required amount of metal from rear face of the camshaft gear hub. Refer to text.

Fig. CS48 — In rare cases, replacement timing gears are not punched with timing marks. When such cases are encountered, transfer timing marks from the old gear to the new, using a straight edge and keystock as shown.

Fig. CS49A—Timing gear train on Diesel models 40 and 50 when equipped with a Bosch model APE injection pump. Notice that the injection pump is chain driven. Model 30 Diesels equipped with an APE pump are similar.

COCKSHUTT 20-30-40-50   Paragraphs 61-64

CS50. Lift tappets off camshaft and hold in the up position with clothes pins as shown. Working through openings in camshaft gear, unbolt the camshaft thrust plate and withdraw camshaft.

The recommended camshaft end play of 0.005-0.009 is controlled by the thickness of the thrust plate. If the end play is excessive, renew the thrust plate.

Check the camshaft against the values which follow:
No. 1 (front) journal
   diameter 1.8715-1.8725
No. 2 (center) journal
   diameter 1.7457-1.7465
No. 3 (rear) journal
   diameter 1.2465-1.2475

When installing the camshaft, reverse the removal procedure and make certain that the valve timing marks are in register as outlined in paragraph 55. Refer to paragraph 201 for method of checking the ignition timing. Adjust the valve tappet gap to 0.014 (hot).

62. CAMSHAFT BEARINGS. To remove the camshaft bearings, first remove the engine as outlined in paragraph 40 and the camshaft as in the preceding paragraph 61. Then, remove clutch, flywheel housing and flywheel. Remove the crankcase rear end plate and extract the soft plug from behind the camshaft rear bearing. Using a suitable puller or drift, remove camshaft bearings; then, clean the bearing oil passages in the cylinder block.

Using a closely piloted arbor, install the bearings so that oil hole in bearings is in register with oil holes in the cylinder block. The camshaft bearing journals should have a normal operating clearance in the bearings of 0.002-0.004 for the front and rear journals, 0.003-0.0045 for the center journal. It is therefore necessary to align ream the bearings after installation to provide the bearing inside diameters which follow:
Camshaft bearing inside diameter.
No. 1 (front) 1.8745-1.8755
No. 2 (center) 1.7495-1.7502
No. 3 (rear) 1.2495-1.2505

When installing the soft plug at rear camshaft bearing, use Permatex or equivalent to obtain a better seal.

Models 30-40-50

The camshaft is supported in four precision steel-backed, babbitt-lined bearings. The shaft journals have a normal operating clearance of 0.002-0.0046 in the bearings. If the journal clearance exceeds 0.0065 the bearings and/or shaft should be renewed. To renew the bearings, follow the procedure outlined in paragraph 64. To renew the camshaft, refer to the following paragraph.

63. CAMSHAFT. To remove the camshaft, first remove the crankcase front cover as outlined in paragraph 57; then, proceed as follows: Remove valve cover, rocker arms assembly and push rods. Insert long dowel pins down through the push rod openings in crankcase and tap the dowels into the tappets (cam followers). Lift dowels up, thereby holding tappets away from camshaft and hold dowels in the raised position by installing clothes pins as shown in Fig. CS51. Remove oil pan and oil pump. Working through openings in camshaft gear, remove the camshaft thrust plate retaining cap screws and withdraw camshaft from engine.

Recommended camshaft end play should be 0.003-0.008 for Diesel engines, 0.002-0.008 for non-Diesels. Refer to timing gears (paragraph 60) for the method of checking and adjusting the camshaft end play.

Check the camshaft against the values which follow:
No. 1 (front) journal
   diameter 1.998-1.999
No. 2 journal diameter 1.998-1.999
No. 3 journal diameter 1.998-1.999
No. 4 (rear) journal
   diameter 1.248-1.249

When installing the camshaft, reverse the removal procedure and make certain that the valve timing marks are in register as outlined in paragraph 55. Caution: Make certain that the drilled camshaft thrust plate retaining cap screw is installed in the lower hole. On non-Diesel models, refer to paragraph 78B for the oil pump installation procedure. Check and adjust the ignition timing as outlined in paragraph 201. Adjust the valve tappet gap to 0.012 (hot).

64. CAMSHAFT BEARINGS. To remove the camshaft bearings, first remove the engine as outlined in paragraph 41 for model 30 or paragraph 42 for models 40 and 50; then, remove camshaft as in the preceding paragraph 63. Remove clutch, flywheel and the engine rear end plate. Extract the soft plug from behind the camshaft rear bearing and remove the bearings.

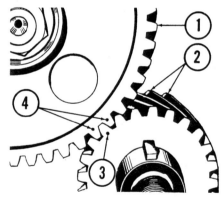

Fig. CS50A—Model 20 valve timing marks. The double punch marked tooth space (4) on the camshaft gear (1) should be meshed with the single punch marked tooth (3) on the crankshaft gear (2).

Fig. CS50—When removing the camshaft on model 20, the valves can be held up with wedges as shown and the tappets can be held in the raised position with spring type clothes pins.

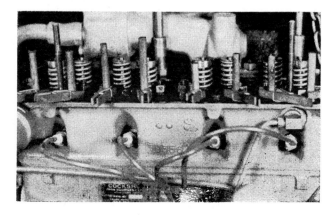

Fig. CS51—When camshaft is being removed on model 30 tractors, the mushroom type tappets (cam followers) can be held in the raised position by tapping dowels into the tappets, then holding the dowels up with spring type clothes pins. A similar procedure can be used on models 40 and 50.

CS-35

An approved method for removing the bearings is to split the bearings with a hack saw blade and drive the bearings out with a chisel. Exercise special care during this operation, however, to avoid damaging the crankcase bearing bores. Clean the camshaft bearing oil holes in crankcase.

Using a closely piloted arbor, install the bearings so that oil hole in bearings is in register with oil holes in the cylinder block.

The inside diameter of the camshaft bearings after installation should be as follows:

No. 1 (front) 2.0010-2.0026
No. 2 .... 2.0010-2.0026
No. 3 .... 2.0010-2.0026
No. 4 (rear) 1.2510-1.2526

Although the camshaft bearings are pre-sized, it is highly recommended that the bearings be checked after installation for localized high spots. The camshaft bearing journals should have a normal operating clearance in the bearings of 0.002-0.0046.

When installing the soft plug at rear camshaft bearing, use Permatex or equivalent to obtain a better seal.

CONNECTING ROD AND PISTON UNITS

Models 20-30-40-50

65. Piston and connecting rod units are removed from above after removing cylinder head and oil pan. The procedure for removing the cylinder head is outlined in paragraph 43 for model 20, paragraph 44 for non-Diesel models 30, 40 and 50 and paragraph 45 for Diesel models 30, 40 and 50.

Cylinder numbers are stamped on the connecting rod and the cap. When reinstalling the rod and piston units, make certain that numbers are in register and face toward the proper side of the engine.

Model 20. Each connecting rod has an oil squirt hole drilled in the side. When reinstalling the rod, the oil squirt hole and the cylinder numbers on rod and cap face toward camshaft side of engine.

Models 30, 40 and 50. Cylinder numbers on rod and cap face opposite the camshaft side of engine.

Tighten the connecting rod bolts to a torque of 35-40 Ft.-Lbs. on Model 20, 50-65 Ft.-Lbs. on non-Diesel models 30, 40 and 50 and 35-55 Ft.-Lbs. on Diesel models 30, 40 and 50.

PISTONS, RINGS AND CYLINDERS (OR SLEEVES)

Model 20

66. Each piston is fitted with four rings; three 1/8 inch wide compression rings and one 1/4 inch wide oil control ring. Check pistons and rings against the values which follow:

Continental F124

Compression ring end
  gap 0.008-0.013
Oil ring end gap....... 0.007-0.017
Compression ring side
  clearance 0.002-0.004
Oil ring side clearance.. 0.001-0.0025

Continental F140

Compression ring end
  gap 0.007-0.017
Oil ring end gap....... 0.008-0.016
Compression ring side
  clearance 0.0015-0.0035
Oil ring side clearance 0.001-0.0025

Standard size cylinder bore is 3.000-3.002 for the F124 engine and 3.1875-3.1895 for the F140 engine. The desired piston skirt to cylinder wall clearance is 0.003. The clearance can be considered satisfactory when a

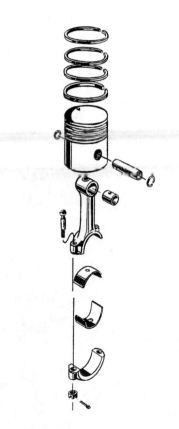

Fig. CS52—Exploded view of model 20 connecting rod, piston and related parts. Rod bolts should be tightened to a torque of 35-40 Ft.-Lbs.

spring scale pull of 5-10 pounds is required to withdraw a 1/2 inch wide, 0.003 thick feeler gage from between piston skirt and cylinder wall.

Cylinders should be rebored if out-of-round and/or if taper is excessive. Pistons are available in oversizes of 0.020, 0.040 and 0.060. Rings are available in oversizes of 0.020, 0.040 and 0.060. Standard size service re-ring sets will accommodate a 0.010 oversize cylinder bore.

Models 30-40-50 (Non-Diesel)

67. Each cast iron piston is fitted with four rings; three 1/8 inch wide compression rings and one 3/16 inch wide oil control ring. Check pistons and rings against the values which follow:

Compression ring end gap 0.010 Min.

Oil ring end gap .... 0.010 Min.

Compression ring side
  clearance (desired) .0.0015-0.0035

Compression ring side
  clearance (Max.) 0.005

Oil ring side
  clearance (desired).. 0.001-0.0025

Oil ring side clearance (Max.) 0.004

Models 30 and 40 are equipped with renewable, wet type sleeves which have a bore diameter of 3.4372-3.4379. Model 50 is not equipped with sleeves and the standard cylinder bore diameter is 3.749-3.751. The cylinders should be rebored or the sleeves should be renewed if the out-of-round exceeds 0.003 and/or taper exceeds 0.009.

The desired clearance between piston skirt and cylinder wall or sleeve is 0.0026-0.0041 for models 30 and 40, 0.0019-0.0047 for model 50. The maximum allowable clearance is 0.008.

Pistons for non-sleeved engines are available in oversizes of 0.020 and 0.040. Piston rings are available for non-sleeved engines in oversizes of 0.020 and 0.040. Two types of rings are available for all models: The regular type is for use in new sleeves and newly rebored cylinder blocks or in sleeves and blocks that are not worn more than 0.004. The service re-ring type is for use in worn sleeves and cylinder blocks.

Refer to paragraph 69 for R&R of sleeves.

Models 30-40-50 (Diesels)

68. Each aluminum piston is fitted with five rings; three 1/8 inch wide compression rings and two 3/16 inch

wide oil control rings. Check pistons and rings against the values which follow:

Models 30 & 40

Compression ring end
  gap 0.009-0.014
Oil ring end gap.......0.009-0.014
Top compression ring side clearance (desired) 0.003-0.005
Top compression ring side clearance (Max.) 0.007
Other compression ring side clearance (desired) 0.002-0.004
Other compression ring side clearance (Max.) 0.0055
Oil ring side clearance (desired) 0.0015-0.0035
Oil ring side clearance (Max.) 0.005

Model 50

Compression ring end
  gap 0.012-0.017
Oil ring end gap.......0.009-0.014
Top compression ring side clearance (desired) 0:003-0.005
Top compression ring side clearance (Max.) 0.007
Other compression ring side clearance (desired) 0.002-0.004
Other compression ring side clearance (Max.) 0.0055
Oil ring side clearance (desired) 0.0015-0.0035
Oil ring side clearance (Max.)..........0.005

Models 30 and 40 are equipped with renewable, wet type sleeves which have a bore diameter of 3.4372-3.4379.

Model 50 is not equipped with sleeves and the cylinders have a bore diameter of 3.749-3.751.

The cylinders should be rebored or the sleeves should be renewed if the out-of-round exceeds 0.003 and/or taper exceeds 0.009.

The desired clearance between piston skirt and cylinder wall or sleeve is 0.004-0.005 for models 30 and 40, 0.0037-0.0065 for model 50. The maximum allowable clearance is 0.0095 for models 30 and 40, 0.010 for model 50.

Piston to sleeve or cylinder wall clearance can be considered satisfactory if a spring scale pull of 2-5 lbs. is required to withdraw a 0.002 thick, ½ inch wide feeler gage.

Pistons for non-sleeved engines are available in oversizes 0.020 and 0.040. Piston rings are available for non-sleeved engines in oversizes of 0.020 and 0.040. Two types of rings are available for all models: The regular type is for use in new sleeves and newly rebored cylinder blocks or in sleeves and blocks that are not worn more than 0.004. The service re-ring type is for use in worn sleeves and cylinder blocks.

Refer to paragraph 69 for R&R of sleeves.

R&R CYLINDER SLEEVES

Models 30-40

69. The wet type cylinder sleeves can be renewed after removing the connecting rod and piston units. Refer to paragraph 65. Coolant leakage at bottom of sleeves is prevented by two rubber "O" rings. The cylinder block is counter-bored at the top to receive the sleeve flange and the head gasket forms the coolant seal at this point. The sleeves can be removed, using a special puller; or, the sleeves can be removed by placing a wood block against bottom edge of sleeve and tapping the block with a hammer.

Before installing new sleeves, thoroughly clean the cylinder block, paying particular attention to the seal seating surfaces at bottom and the counterbore at top. All sleeves should enter crankcase bores full depth and should be free to rotate by hand when tried in bores without "O" rings. After making a trial installation without "O" rings, remove the sleeves and install the "O" rings making certain that the "O" rings are not twisted. To facilitate installation of the sleeves and to keep from cramping or causing the "O" rings to bulge, coat the "O" rings with hydraulic brake fluid or a thick soap solution.

PISTON PINS

Models 20-30-40-50

70. The full floating type piston pins are retained in the piston pin bosses by snap rings. Piston pins are available in oversizes of 0.003, 0.005 and 0.010 for model 20, 0.005 and 0.010 for models 30, 40 and 50. Check the piston pin against the values which follow:
Standard piston pin diameter

  Model 20 ....0.8591 -0.8593
  Models 30, 40 and 50 0.99955-0.99975
Piston pin clearance in piston
  Model 20 ....0.0001T -0.0003L
  Models 30, 40 and
    50 ....0.00025T-0.00045L
Piston pin clearance in rod bushing
  Model 20 ....0.0002 -0.0006
  Models 30, 40 and 50 0.00015-0.00085

CONNECTING RODS AND BEARINGS

Models 20-30-40-50

71. Connecting rod bearings in model 20 and non-Diesel models 30, 40 and 50 are of the renewable, steel-backed, babbitt-lined, slip-in type. Connecting rod bearings in Diesel models 30, 40 and 50 are of the renewable, steel-backed, copper-lead-lined, slip-in type. The bearings can be renewed after removing oil pan and bearing caps. When installing new bearing shells, make certain that the bearing shell projections engage the milled slot in connecting rod and bearing cap and that cylinder numbers on the rod and cap are in register and face toward camshaft side of engine on model 20, opposite to camshaft on models 30, 40 and 50. Bearing inserts are available in undersizes of 0.002,

Fig. CS53—Exploded view of non-Diesel models 30, 40 and 50 connecting rod, piston and associated parts. Rod bolts should be tightened to a torque of 50-65 Ft.-Lbs.

0.010, 0.020 and 0.040 for model 20, 0.002, 0.010, 0.012, 0.020 and 0.040 for models 30, 40 and 50. Check the crankshaft crankpins and the bearing inserts against the values which follow:

Crankpin diameter (Standard)
   Model 20 . . . . 1.9365-1.9375
   Models 30, 40 and 50 . . . 1.9975-1.9985
Rod bearing running clearance
   Model 20 . . . . 0.0002-0.0022
   Models 30, 40 and 50
    (desired) 0.0015-0.0035
   Models 30, 40 and 50
    (Max.) 0.006
Rod side play
   Model 20 . . . . 0.006-0.010
   Models 30, 40 and 50
    (desired) 0.003-0.009
   Models 30, 40 and 50
    (Max.) 0.013
Rod bolt torque (Ft.-Lbs.)
   Model 20 35-40
   Models 30, 40 and 50
    (non-Diesels) 50-65
   Models 30, 40 and 50 (Diesels) 35-55

CRANKSHAFT AND MAIN BEARINGS

Model 20

72. Crankshaft is supported in three steel-backed, babbitt-lined precision type bearings which can be renewed after removing oil pan, front and rear filler blocks and main bearing caps. Normal crankshaft end play of 0.004-0.006 is controlled by two thrust washers (5—Fig. CS55) and a series of 0.002 and 0.008 thick shims (6). The end play can be adjusted by varying the number of shims after crankshaft gear is removed as outlined in paragraph 58.

To remove the crankshaft, first remove engine as outlined in paragraph 40; then, remove clutch, flywheel housing, flywheel and engine rear end plate. Remove the crankcase front cover (timing gear cover), camshaft gear and engine front end plate. Remove oil pan, front and rear filler blocks, and rod and main bearing caps. Lift crankshaft from engine.

Check crankshaft and main bearings against the values which follow:
Crankpin diameter
   (Standard) 1.9365-1.9375
Main journal diameter
   (Standard) . . . . 2.249 -2.250
Main bearing running
   clearance 0.0002-0.0024
Main bearing bolt torque
   (Ft. Lbs.) 85-95
Main bearing inserts are available in undersizes of 0.002, 0.010, 0.020 and 0.040.

Models 30-40-50

73. The crankshaft is supported in three main bearings on model 30 tractors and seven main bearings on 40 and 50 models. Main bearings are of the steel-backed, babbitt-lined, slip-in precision type on non-Diesels; steel-backed, copper-lead-lined, slip-in precision type on Diesels. Main bearings can be renewed after removing oil pan and main bearing caps. Desired crankshaft end play is 0.002-0.007 for non-Diesels, 0.002-0.006 for Diesels. Maximum allowable end play is 0.011 for all models. Crankshaft end play is corrected by renewing the center main bearing shells.

To remove the crankshaft, first remove the engine as outlined in paragraph 41 for model 30 and paragraph 42 for models 40 and 50. Remove clutch, flywheel and engine rear end plate. Remove valve cover, rocker arms assembly and push rods. Remove oil pan, oil pump, and rod and main bearing caps. Remove crankcase front cover (timing gear cover), unbolt camshaft thrust plate, withdraw camshaft and remove the engine front end plate. Lift crankshaft from engine.

Check crankshaft and main bearings against the values which follow:
Crankpin diameter (Standard) 1.9975-1.9985
Main journal diameter (Standard) 2.497-2.498
Main bearing running clearance (Diesels) (desired) . . 0.0023-0.0045
Main bearing running clearance (Diesels) (Max.) 0.0065
Main bearing running clearance (non-Diesels) (desired) 0.0015-0.0035
Main bearing running clearance (non-Diesels) (Max.) 0.0065
Main bearing bolt torque (Diesels) (Ft. Lbs.) 125-135
Main bearing bolt torque (non-Diesels) (Ft. Lbs.) 115-135
Main bearings are available in undersizes of 0.002, 0.010, 0.012, 0.020 and 0.040.

CRANKSHAFT REAR OIL SEAL

Model 20

74. The crankshaft rear oil seal is of two piece jute. The upper half is retained in an oil guard and the lower half in the rear filler block. To renew the oil seal, remove oil pan and rear filler block (22—Fig. CS56). The lower half of the seal can be renewed at this time. Using a dowel or soft drift, tap on one end of the oil guard (20). After the oil guard has moved a slight amount, the guard can be "rolled" out of the groove in the crankcase. Install oil seal by reversing the removal procedure.

Models 30-40-50

75. The crankshaft rear oil seal is of the one piece spring loaded type. The seal is retained in the oil seal retainer plate (15—Fig. CS57) which is retained to the front face of the engine rear end plate by six screws. Circular seal (13) fits around seal retainer (15) and forms the rear seal for the oil pan. To remove the spring loaded oil seal (14), first remove the flywheel as outlined in paragraph 77 and pry the seal out of the groove in the seal retainer

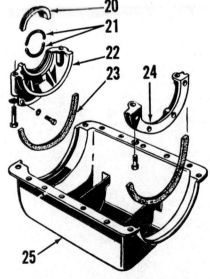

Fig. CS56—Model 20 oil pan, filler blocks and rear main bearing oil seal.
20. Oil guard
21. Jute seal
22. Rear filler block
23. Cork seal
24. Front filler block
25. Oil pan

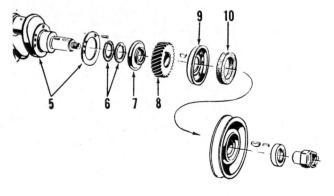

Fig. CS55—Crankshaft end play on model 20 tractors is controlled by thrust washers (5) and can be adjusted by varying the number of shims (6).
5. Thrust washers
6. Shims
7. Thrust plate
8. Crankshaft gear
9. Oil slinger
10. Pulley seal

COCKSHUTT 20-30-40-50

plate. To remove the circular seal (13), it is necessary to support rear of engine and remove the engine rear end plate.

OIL PAN

Models 20-30-40-50

76. The procedure for removing and reinstalling the model 20 oil pan is evident. When installing the oil pan on models 30, 40 and 50, the following should be observed.

Five of the oil pan retaining cap screws are longer than the others. These longer cap screws pass through the crankcase front cover (timing gear cover) and into the front face of the oil pan. The bottom three of these cap screws must be fitted with copper washers. When installing the oil pan, install all of the cap screws loosely; then, tighten the front five cap screws before tightening any of the others.

FLYWHEEL

Models 20-30-40-50

77. To remove the flywheel on model 20, first remove the engine as outlined in paragraph 40. Then, remove clutch, flywheel housing and flywheel. On models 30, 40 and 50, the flywheel can be removed after removing the clutch. The starter ring gear can be removed by drilling and splitting same with a cold chisel. To facilitate installation, boil the new gear in oil for fifteen minutes or heat the ring gear evenly with a torch. Install ring gear with beveled end of teeth facing rearward on models 20 and 30, forward on models 40 and 50.

OIL PUMP

Models 20-30-40-50

78. The oil pump, which is gear driven from the camshaft, can be removed after removing the oil pan.

To disassemble the pump, proceed as follows: File off the head of the pin which retains the pump driving gear to the pump drive shaft and using a small punch as shown in Fig. CS60, remove the pin. Remove cover from pump housing and withdraw drive shaft and gear and idler gear and shaft from pump housing. To remove the driver gear from the drive shaft, press gear further up the shaft until the snap ring is exposed. Remove the snap ring and press the gear off the shaft. On model 20, check the pump drive shaft and the drive shaft bushing (3—Fig. CS61) in pump housing for wear. If the clearance between the drive shaft and the housing is excessive, renew the bushing. On models 30, 40 and 50, inspect the oil

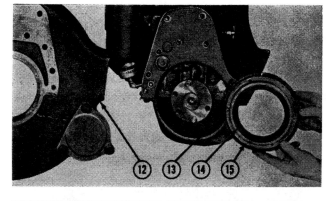

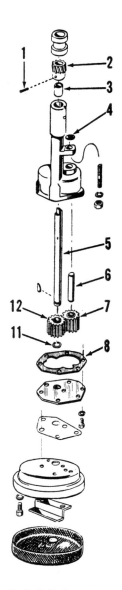

Fig. CS57—Models 30, 40 and 50 crankshaft rear oil seal (14) is of the spring loaded type and is carried in seal retainer (15) which is bolted to the engine rear end plate (12). Circular seal (13) forms the rear seal for the oil pan.

Fig. CS60 — On all models, the oil pump drive gear is retained to the drive shaft by a pin. After grinding off the end of the pin, the pin can be drifted out as shown.

Fig. CS61—Exploded view of model 20 oil pump. When installing the oil pump, make certain that spacer washer (4) is installed between the pump bolting flange and the cylinder block.

1. Pin
2. Gear
3. Bushing
4. Spacer washer
5. Drive shaft
6. Idler gear shaft
7. Idler gear
8. Gasket
11. Snap ring
12. Drive gear

seal (9—Fig. CS62) in the pump body. There is no need to remove the seal unless the seal is damaged. If, however, the seal is removed, always install a new seal.

On models 30, 40 and 50, to prevent possible damage to the drive shaft oil seal, install the pump drive shaft as follows: With the oil seal in place, temporarily install the idler shaft in the pump body at oil seal end of pump. With the idler shaft extending through the oil seal, insert the drive shaft into the pump body, from drive end, until drive shaft butts against the idler shaft. Press drive shaft into position, forcing the idler shaft out of the pump body. The remainder of the assembly procedure is evident.

78A. When installing the oil pump on model 20, make certain that the 1/8 inch thick spacer washer (4—Fig. CS61) is installed between the cylinder block and the pump bolting flange; otherwise, the pump drive gear and the camshaft gear will not mesh properly. Check and retime the ignition distributor as outlined in paragraph 201.

78B. On non-Diesel models 30, 40 and 50, install the oil pump as follows: Crank engine until No. 1 piston is coming up on compression stroke and "STATIC SPARK" mark on model 30 flywheel or "T.D.C." mark on model 40 or 50 flywheel is in the center of the timing hole in rear mounting plate as shown in Fig. CS65. Remove the distributor cap and place the rotor arm in the number one firing position. Install oil pump so that drive slots mesh properly. The oil pump drive gear on Diesel models can be meshed in any position.

On models 30, 40 and 50, when the oil pump is completely in place, there will be a space of about 3/16 inch between the crankcase and the pump bolting flange. CAUTION: Do not tighten the pump retaining cap screws too tight, or the flange ears might break off.

OIL PRESSURE RELIEF VALVE

Model 20

79. Normal oil pressure of 20-25 psi is adjustable and controlled by a spring-loaded oil pressure relief valve which is located on right side of engine. The procedure for removing the relief valve and spring is evident. To adjust the oil pressure, vary the number of shim washers (33—Fig. CS67).

Models 30-40-50

80. Normal oil pressure of 20-25 psi is controlled by a spring-loaded oil pressure relief valve. The pressure is adjusted by a slotted screw (26—Fig. CS68) on left front side of crankcase. If difficulty is encountered when attempting to regulate the oil pressure, check the drilled cap screw which holds the timing gear thrust plate in place. If the screw is too long, it will interfere with the valve action. The drilled cap screw dimensions are 7/16-14 x 9/16 inch.

CARBURETOR (Not L.P.G.)

Models 20-30-40-50

100. Gasoline and distillate carburetors are either Marvel-Schebler or Zenith and their applications are as follows:

Tractor Model	Carburetor Make	Carburetor Model
20 Gas.	M.S.	TSX-520
20 Dis.	M.S.	TSX-531
30 Gas.	Zen.	161JX7-10432A
30 Gas.	M.S.	TSX-264
30 Dis.	Zen.	161X7-11080
40 Gas.	Zen.	162J9-10946A
40 Dis.	Zen.	162J9-11489
50 Gas.	Zen.	162J9-11509

Refer to Carburetor Section of STANDARD UNITS for adjustment, circuit description and calibration data.

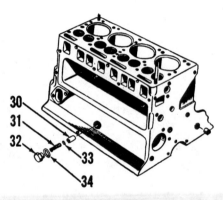

Fig. CS67—Model 20 oil pressure relief valve exploded from right side of cylinder block.

30. Oil pressure relief valve
31. Relief valve spring
32. Plug
33. Adjusting washers
34. Gasket

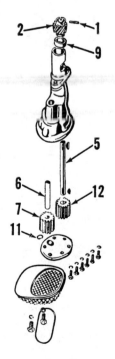

Fig. CS62—Exploded view of a typical oil pump as used on models 30, 40 and 50.

1. Pin
2. Gear
5. Drive shaft
6. Idler shaft
7. Idler gear
9. Oil seal
11. Snap ring
12. Drive gear

Fig. CS65—Static spark mark on flywheel of model 30 non-Diesel tractor. The equivalent mark on models 40 and 50 is "T.D.C."

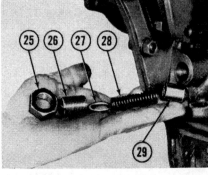

Fig. CS68—Models 30, 40 and 50 oil pressure relief valve exploded from left front side of cylinder block.

25. Nut
26. Pressure adjusting screw
27. Gasket
28. Relief valve spring
29. Relief valve

L.P. GAS SYSTEM

Models 30 and 40 tractors are available with an LP-gas system using Century equipment. The systems are designed to operate with the fuel supply tank not more than 80% filled.

The LP-gas systems include a model 3C-705 carburetor on model 40 and a 3C-704 carburetor on model 30. The systems also include an M4C converter (regulator) equipped with a solenoid operated primer, a combined gas filter and fuelock (solenoid operated fuel shutoff) and a fuel tank.

CARBURETOR

Models 30-40

The carburetor has 2 points of mixture adjustment plus a throttle stop screw. Refer to Fig. CS75. This carburetor is designed for the engine to start with throttle valve in fully closed position.

105. THROTTLE STOP SCREW. The throttle stop screw on the carburetor throttle should be adjusted to provide an engine slow idle speed of 400-450 rpm.

106. IDLE MIXTURE. This adjustment synchronizes the carburetor throttle valve (9—Fig. CS75) to the gas valve (10). To make the adjustment, proceed as follows: With the engine warm, place the governor hand control lever in the idle position. Remove the cotter pin from one end of the carburetor drag link (7). Rotate either drag link adjusting screw (A or B) **in** or **out** until the engine runs smoothly. Turning either adjusting screw **in** will enrich the mixture. Turn the throttle stop screw out to obtain the desired engine idle speed of 400-450 rpm. Readjust the drag link adjusting screw for the fastest rpm; or, if a vacuum gage is available, remove the pipe plug from the intake manifold, connect the gage and adjust the drag link adjusting screw to obtain the highest steady vacuum. If the engine still idles too fast, reset the throttle stop screw and readjust the drag link screw. After the adjustment is complete, lock the adjustment by inserting a cotter pin in the drag link.

107. POWER ADJUSTMENT. The power adjustment is controlled by the position of the carburetor spray bar (P—Fig. CS77). The spray bar has no effect on the mixture except at wide open throttle.

108. CENTURY METHOD WITHOUT ANALYZER. The spray bar is factory adjusted so that the holes in the bar are at approximately a 45 degree angle. With the engine running, loosen the spray bar set screw and rotate the spray bar (P) until the arrow (stamped on end of spray bar) is toward the "L" (lean) mark stamped on carburetor body. Open throttle valve quickly to the wide open position. If engine falters, rotate spray bar toward the "R" (rich) mark. Repeat the acceleration test until the engine accelerates smoothly. Tighten the spray bar set screw to lock the adjustment.

LP-GAS FILTER AND FUELOCK

Models 30-40

110. The combined filter and fuelock are mounted as shown in Figs. CS78 or 79. The fuelock is a solenoid operated fuel shut-off valve which operates on the ignition circuit.

111. In most cases, the fuel strainer can be cleaned by opening the drain cock (located on lower side of filter) and then opening the fuel tank liquid valve to blow out any accumulation of foreign material.

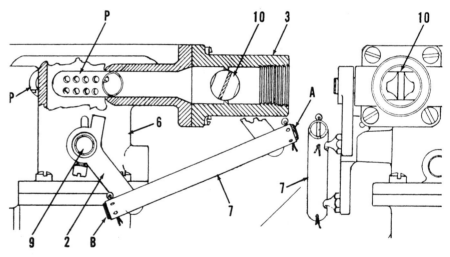

Fig. CS77—Sectional view of Century LP-Gas carburetor, showing spray bar (P) and gas valve (10).

P. Spray bar
2. Throttle valve lever
3. Gas valve body
6. Throttle body
7. Drag link
9. Throttle valve
10. Gas valve

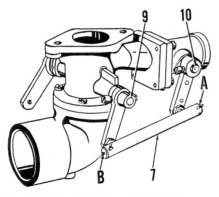

Fig. CS75—Models 30 and 40 LP-Gas carburetor, showing points of adjustment.

A & B. Adjusting screws
9. Throttle valve
10. Gas valve

Fig. CS78—Model 30 tractor, showing the installation of the Century LP-Gas equipment.

Paragraphs 112-113

COCKSHUTT 20-30-40-50

LP-GAS CONVERTER
Models 30-40

112. HOW IT OPERATES. Refer to CS80. Fuel from the supply tank passes through the fuel filter and the solenoid operated fuelock valve and enters the converter inlet (13) at a tank pressure of 20 to 175 psi and into the surge chamber where it is reduced from tank pressure to about 6 psi. Flow through the fuel inlet valve is controlled by the inlet pressure diaphragm (36). When the fuel enters the surge chamber, it expands rapidly and is converted from a liquid to a gas by heat from the engine cooling system. The vaporized fuel then passes through the low pressure valve (43) and in to the low pressure chamber where it is drawn off at a pressure slightly below atmospheric. The low pressure valve is controlled by a large diaphragm (30) and spring (44).

A balance line (5) is connected to the carburetor air inlet horn so as to reduce the flow of fuel and thus prevent over richening of the mixture which would otherwise result when the air cleaner or the air inlet system becomes restricted.

A solenoid operated primer (23), which opens the fuel outlet valve, is mounted on the face of the unit. The primer is used for starting purposes to fill the lines when the throttle is closed. The primer, which can be operated manually, is actuated by a separate switch located on the instrument panel.

113. CONVERTER OVERHAUL. Remove the unit from the tractor and completely disassemble, using Fig. CS80 as a reference. Thoroughly wash all parts and blow out all passages with compressed air. Inspect each part carefully and renew any which are worn.

Fig. CS79—Century fuel filter, fuelock and converter installation on model 40

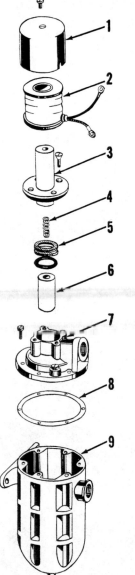

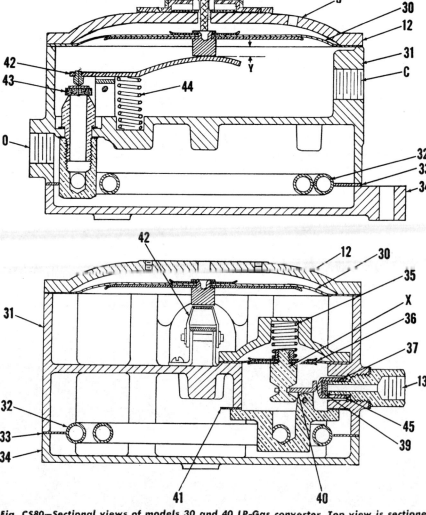

Fig. CS79A—Exploded view of Century filter and fuelock for models 30 and 40.

1. Fuelock case
2. Coil
3. Plunger
4. Spring
5. Gasket set
6. Plunger
7. Base and filter pack
8.
9. Body

Fig. CS80—Sectional views of models 30 and 40 LP-Gas converter. Top view is sectioned through the fuel outlet valve (low pressure) section. Lower view is sectioned through the fuel inlet valve (high pressure) section.

C. Fuel outlet
0. Water outlet
5. Balance line connection
12. Cover
13. Fuel inlet
23. Solenoid
30. Outlet valve diaphragm
31. Regulator body
32. Vaporizer coil
33. Gasket
34. Converter base
35. Spring
36. Inlet valve diaphragm
37. Inlet valve
39. Inlet valve seat
40. Valve lever
42. Valve lever
43. Outlet valve
44. Spring
45. Spring

CS-42

COCKSHUTT 20-30-40-50

Paragraphs 114-132

114. Before reassembling the unit, check the inlet pressure diaphragm for proper adjustment as follows: With the fuel inlet valve closed, the shoulder (X—Fig. CS80) of the diaphragm lever link should be level or not more than 1/16 inch above the machined face of the main casting for the inlet pressure diaphragm. To obtain this adjustment, bend the diaphragm lever (40) at the valve contacting end.

Continue to reassemble the unit, and before installing the fuel outlet diaphragm (30) and cover (12), note dimension (Y) which is measured from the face of the low pressure side of the casting to the diaphragm contacting surface of the low pressure valve operating lever (42) when the valve is in the closed position. Dimension (Y) should be 1/8-5/32 inch and can be obtained by bending the fuel outlet valve lever.

115. After installing the converter, bleed the air from the converter water jacket by opening the converter drain cock and allowing water to run until there is a continuous flow without air bubbles.

SYSTEM TROUBLE SHOOTING
Models 30-40

116. SYMPTOM. Sweating or frosting fuel filter.

CAUSE AND CORRECTION. *Fuel filter is clogged, thereby restricting the flow of fuel. In most cases, the strainer can be cleaned by opening the drain cock located on the lower side of the filter bowl; then, opening the fuel tank liquid valve so as to blow the foreign material out of the filter.*

117. SYMPTOM. Converter shows moisture and frost after standing.

CAUSE AND CORRECTION. *Trouble is due to leaking valves.*

The fuel inlet valve can be cleaned and serviced as follows: Disconnect fuel filter converter line and remove the high pressure valve and body. The valve seat (39—Fig. CS80) can be removed and turned over so that the unused portion of the seat contacts the shoulder of the fuel inlet fitting.

The fuel outlet valve can be cleaned as follows: First remove the converter cover (12). Open the fuel tank liquid valve and press down on the valve lever (42) to allow liquid fuel to wash over the seat.

Check the valve levers as per paragraph 114.

118. SYMPTOM. Converter shows frost when engine is hot.

CAUSE AND CORRECTION. *Trouble is caused by poor water circulation through the converter. To correct the trouble, open the converter drain cock and allow water to run until there is a continuous flow without air bubbles. Also, check the water supply lines for restrictions.*

119. SYMPTOM. Engine fails to start.

CAUSE AND CORRECTION. *Trouble is caused by a restricted fuel filter, inoperative fuelock (shut-off) valve, an overprimed engine or a low fuel tank.*

Check and clean the fuel filter as per paragraph 111.

The fuelock (shut-off) valve can be checked by turning the ignition switch to the on position and listening for the ping as the plunger valve is unseated. If the plunger valve ping cannot be heard, check the electrical circuit.

DIESEL SYSTEM

Models 30-40-50

The Diesel fuel system consists of three basic units; the fuel filters, injection pump and injection nozzles. When servicing any unit associated with the fuel system, the maintenance of absolute cleanliness is of utmost importance. Of equal importance is the avoidance of nicks or burrs on any of the working parts.

Probably the most important precaution that service personnel can impart to owners of Diesel powered tractors, is to urge them to use an approved fuel that is absolutely clean and free from foreign material. Extra precaution should be taken to make certain that no water enters the fuel storage tanks. This last precaution is based on the fact that all Diesel fuels contain some sulphur. When water is mixed with sulphur, sulphuric acid is formed and the acid will quickly erode the closely fitting parts of the injection pump and nozzles.

130. QUICK CHECKS—UNITS ON TRACTOR. If the Diesel engine does not start or does not run properly, and the Diesel fuel system is suspected as the source of trouble, refer to the Diesel System Trouble Shooting Chart and locate points which require further checking. Many of the chart items are self-explanatory; however, if the difficulty points to the fuel filters, injection nozzles and/or injection pump, refer to the appropriate paragraphs which follow:

FUEL FILTERS
Models 30-40-50

The fuel filtering system on early production tractors consisted of a glass sediment bowl, a primary filter and a secondary filter. Each filter contained a renewable type element. The filtering system on late production tractors consists of a glass sediment bowl, a primary filter of the replaceable element type, a secondary filter of the replaceable element type and a final stage filter of the sealed type.

The Cockshutt factory recommends installing a final stage filter assembly on all models which were not originally equipped with the three stage system. Due to this recommendation, only the late production three stage system will be covered in this manual.

132. CIRCUIT DESCRIPTION AND MAINTENANCE. Fuel from the fuel tank flows through a glass sediment bowl (A—Fig. CS100) which should be removed, drained and cleaned each day, prior to starting the engine. The fuel then flows to the renewable element type primary filter (B). The drain cock at the bottom of the primary filter should be opened and a small quantity of fuel drained each day, prior to starting the engine. From the primary filter, the fuel passes through the transfer pump (C—Fig. CS 101) to the renewable element type secondary filter (E). If any signs of water are apparent when draining fuel from the primary filter, a small quantity of fuel must be drained from the secondary filter. The by-pass valve (D) which is located on the inlet side of the secondary filter allows

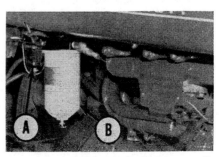

Fig. CS100—Model 30 Diesel glass sediment bowl (A) and the first stage, renewable element type fuel filter (B). Models 40 and 50 Diesels are similar.

CS-43

DIESEL SYSTEM TROUBLE SHOOTING CHART

	Sudden Stopping of Engine	Lack of Power	Engine Hard to Start	Irregular Engine Operation	Engine Knocks	Excessive Smoking	Excessive Fuel Consumption
Lack of fuel	★	★	★	★			
Water or dirt in fuel	★	★	★	★			
Clogged fuel lines	★	★	★	★			
Inferior fuel	★	★	★	★			
Faulty transfer pump	★	★	★	★			
Faulty injection pump timing		★	★	★	★	★	★
Air traps in system	★	★	★	★			
Clogged fuel filters		★	★	★			
Deteriorated fuel lines	★						★
Air leak in suction line	★						
Faulty nozzle				★	★	★	★
Sticking pump plunger		★		★			
Binding pump control rod				★			
Weak or broken governor springs				★			
Weak or broken sump overflow valve spring (APE)		★	★	★			
Fuel delivery valve not seating properly				★			
Weak or broken transfer pump plunger spring		★	★				
No oil in governor (APE)				★			
Improperly set smoke stop		★				★	
Broken spring in by-pass valve	★						

all fuel to be returned to the primary filter for refiltering, EXCEPT the amount of fuel required by the engine, plus the fuel passing through the pump sump bleed line which is required to bleed the system. The fuel then passes through the third or final stage sealed-type filter (F) and into the injection pump sump. The final stage filter must be renewed as a unit.

The fuel transfer pump, pumps fuel through the secondary and final stage filters under a normal pressure of 30 p.s.i. If the secondary and final stage filters become clogged, fuel from the transfer pump will be by-passed to the primary filter and the engine will die from lack of fuel.

132A. To check the filters and/or the transfer pump, install a suitable pressure gage in series between the final fuel filter outlet and the injection pump fuel inlet. Start the engine and observe the pressure gage reading. The gage should register at least 5 p.s.i. If the gage reading is low, renew

Fig. CS101—Models 40 and 50 Diesel fuel filters and model PSB Bosch injection pump installation. Model 30 is similar.

C. Transfer pump
D. By-pass valve
E. Secondary filter
F. Final stage filter

the second stage element and recheck pressure reading. If the gage reading is still low, renew the complete third or final stage filter and recheck the pressure reading. If the gage reading is still low, the transfer pump is not operating properly and same should be renewed and/or overhauled. Refer to paragraphs 154 or 162.

INJECTION NOZZLES
Models 30-40-50

WARNING: *Fuel leaves the injection nozzles with sufficient force (2000 p.s.i.) to penetrate the skin. When testing, keep your person clear of the nozzle spray.*

133. **TESTING AND LOCATING FAULTY NOZZLE.** If the engine does not run properly, or not at all, and the quick checks outlined in paragraph 130 point to a faulty injection nozzle, locate the faulty nozzle as follows:

If one engine cylinder is misfiring, it is reasonable to suspect a faulty nozzle. Generally, a faulty nozzle can be located by loosening the high pressure line fitting on each nozzle holder in turn, thereby allowing fuel to escape at the union rather than enter the cylinder. As in checking spark plugs in a spark ignition engine, the faulty nozzle is the one which, when its line is loosened, least affects the running of the engine.

134. Remove the suspected nozzle from the engine as outlined in paragraph 139. If a suitable nozzle tester is available, check the nozzle as in paragraph 135, 136, 137 and 138. If a nozzle tester is not available, reconnect the fuel line and with the nozzle tip directed where it will do no harm, crank the engine with the starting motor and observe the nozzle spray pattern as shown in Fig. CS103.

If the spray pattern is ragged, as shown in the left hand view, the nozzle valve is not seating properly and same should be reconditioned as outlined in paragraph 140. If cleaning and/or nozzle and tip renewal does not restore the unit and a nozzle tester is not available for further checking, send the complete nozzle and holder assembly to an official Diesel service station for overhaul.

135. **NOZZLE TESTER.** A complete job of testing and adjusting the nozzle requires the use of a special tester such as the Buda Hydraulic Nozzle Tester which is available through the Cockshutt Parts Department under the number of TO-5638, the American Bosch Nozzle Tester TSE 7722D which is available through any of the Bosch authorized service agencies, etc. The nozzle should be tested for leakage, spray pattern and opening pressure.

Operate the tester lever until oil flows and attach the nozzle and holder assembly.

Note: Only clean, approved testing oil should be used in the tester tank.

Close the tester valve and apply a few quick strokes to the lever. If undue pressure is required to operate the lever, the nozzle valve is plugged and same should be serviced as in paragraph 140.

136. *Leakage.* The nozzle valve should not leak at a pressure less than 1700 p.s.i. To check for leakage, actuate the tester handle slowly and as the gage needle approaches 1700 p.s.i., observe the nozzle tip for drops of fuel. If drops of fuel collect at pressures less than 1700 p.s.i., the nozzle valve is not seating properly and same should be serviced as in paragraph 140.

137. *Spray Pattern.* Operate the tester handle at approximately 100 strokes per minute and observe the spray pattern as shown in Fig. CS103. If the nozzle has a ragged spray pattern as shown in the left view, the nozzle valve should be serviced as in paragraph 140.

138. *Opening Pressure.* While operating the tester handle, observe the gage pressure at which the spray occurs. The gage pressure should be 2000 p.s.i. If the pressure is not as specified, remove the nozzle protecting cap, exposing the pressure adjusting screw and locknut. Loosen the locknut and turn the adjusting screw as shown in Fig. CS105 either way as required to obtain an opening pressure of 2000 psi. Note: If a new pressure spring has been installed in the nozzle holder, adjust the opening pressure to 2020 psi. Tighten the locknut and install the protecting cap when adjustment is complete.

139. **REMOVE AND REINSTALL.** Before loosening any lines, wash the nozzle holder and connections with clean Diesel fuel or kerosene. After disconnecting the high pressure and leak-off lines, cover open ends of connections with tape or composition caps to prevent the entrance of dirt or other foreign material. Remove the nozzle holder stud nuts and carefully withdraw the nozzle from cylinder head, being careful not to strike the tip end of the nozzle against any hard surface.

Thoroughly clean the nozzle recess in the cylinder head before reinserting the nozzle and holder assembly. It is important that the seating surfaces of recess be free of even the smallest particle of carbon which could cause the unit to be cocked and result in blowby of hot gases. No hard or sharp tools should be used for cleaning. A piece of wood dowel or brass stock properly shaped is very effective. Do not reuse the copper ring gasket (1—Fig. CS107), always install a new one. Tighten the nozzle holder stud nuts to a torque of 14-16 Ft.-Lbs.

140. **MINOR OVERHAUL OF NOZZLE VALVE AND BODY.** Hard or sharp tools, emery cloth, crocus cloth, grinding compounds or abrasives of any kind should NEVER be used in the cleaning of nozzles. A nozzle cleaning and maintenance kit is available through any American Bosch Service Agency under the number of TSE 7779.

Wipe all dirt and loose carbon from the nozzle and holder assembly with a clean, lint free cloth. Carefully clamp

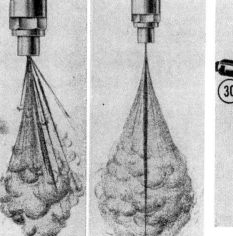

Fig. CS103—Typical spray patterns of a throttling type pintle nozzle. Left: Poor Spray pattern. Right: Ideal spray pattern.

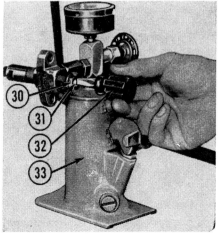

Fig. CS105—Adjusting models 30, 40 and 50 nozzle opening pressure, using a Buda nozzle tester.

30. Nut
31. Adjusting screw
32. Screw driver
33. Nozzle tester

Fig. CS106—Removing models 40 and 50 injection nozzle. Model 30 is similar. The injection pump is a Bosch model APE.

nozzle holder assembly in a soft jawed vise and remove the nozzle holder nut and spray nozzle. Reinstall the holder nut to protect the lapped end of the holder body. Normally, the nozzle valve (V—Fig. CS109) can be easily withdrawn from the nozzle body. If the valve cannot be easily withdrawn, soak the assembly in fuel oil, acetone, carbon tetrachloride or similar carbon solvent to facilitate removal. Be careful not to permit the valve or body to come in contact with any hard surface.

Clean the nozzle valve with mutton tallow used on a soft, lint free cloth or pad. The valve may be held by its stem in a revolving chuck during this cleaning operation. A piece of soft wood well soaked in oil will be helpful in removing carbon deposits from the valve.

The inside of the nozzle body (tip) can be cleaned by forming a piece of soft wood to a point which will correspond to the angle of the nozzle valve seat. The wood should be well soaked in oil. The orifice of the tip can be cleaned with a wood splinter. The outer surfaces of the nozzle body should be cleaned with a brass wire brush and a soft, lint free cloth soaked in a suitable carbon solvent.

Thoroughly wash the nozzle valve and body in clean Diesel fuel and clean the pintle and its seat as follows: Hold the valve at the stem end only and using light oil as a lubricant, rotate the valve back and forth in the body. Some time may be required in removing the particles of dirt from the pintle valve; however, abrasive materials should never be used in the cleaning process.

Test the fit of the nozzle valve in the nozzle body as follows: Hold the body at a 45 degree angle and start the valve in the body. The valve should slide slowly into the body under its own weight. Note: Dirt particles, too small to be seen by the naked eye, will restrict the valve action. If the valve sticks, and it is known to be clean, free-up the valve by working the valve in the body with mutton tallow.

Before reassembling, thoroughly rinse all parts in clean Diesel fuel and make certain that all carbon is removed from the nozzle holder nut. Install nozzle body and holder nut, making certain that the valve stem is located in the hole of the holder body. It is essential that the nozzle be perfectly centered in the holder nut. A centering sleeve is supplied in American Bosch kit TSE 7779 for this purpose. Slide the sleeve over the nozzle with the tapered end centering in the holder nut. Tighten the holder nut, making certain that the sleeve is free while tightening. Refer to Fig. CS110.

Test the nozzle for spray pattern and leakage as in paragraphs 136 and 137. If the nozzle does not leak under 1700 p.s.i, and if the spray pattern is symmetrical as shown in right hand view of Fig. CS103, the nozzle is ready for use. If the nozzle will not pass the leakage and spray pattern tests, renew the nozzle valve and seat, which are available only in a matched set; or, send the nozzle and holder assembly to an official Diesel service station for a complete overhaul which includes reseating the nozzle valve pintle and seat.

150. OVERHAUL OF NOZZLE HOLDER. (Refer to Fig. CS 112.) Remove cap nut (1) and gasket. Loosen jam nut (2) and adjusting screw (3). Remove the spring retaining nut (4) and withdraw the spindle (5) and spring (6). Thoroughly wash all parts in clean Diesel fuel and examine the end of the spindle which contacts the nozzle valve stem for any irregularities. If the contact surface is pitted or rough, renew the spindle. Examine spring seat (7) for tightness to spindle

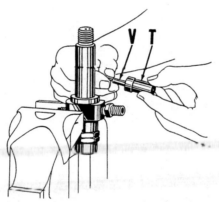

Fig. CS107—Sectional view showing the injection nozzle installation on models 30, 40 and 50. Whenever the nozzle has been removed, always renew the copper gasket (1).

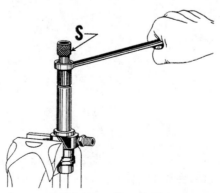

Fig. CS109 — Removing injection nozzle valve (V) from tip (T). If the valve is difficult to remove, soak the assembly in a suitable carbon solvent.

Fig. CS110—Using Bosch tool (S) to center the nozzle tip while tightening the cap nut.

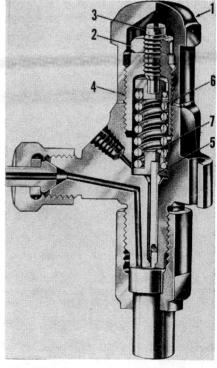

Fig. CS112—Injection nozzle sectional view.
1. Cap nut
2. Jam nut
3. Adjusting screw
4. Spring retaining cap nut
5. Spindle
6. Spring
7. Lower spring seat

and for cracks or worn spots. Renew the spring seat and spindle unit if the condition of either is questionable. Renew any other questionable parts.

Reassemble the nozzle holder and leave the adjusting screw locknut loose until after the nozzle opening pressure has been adjusted as outlined in paragraph 138.

INJECTION PUMP

Models 30, 40 and 50 tractors may be equipped with the following American Bosch injection pumps: The multiple plunger model APE, shown in Fig. CS115; the single plunger models PSB series "B" or "C", shown in Fig. CS116; or the single plunger model PSB series "Y", shown in Fig. CS117. It will be noted that the PSB series "Y" has an externally located governor spring.

The subsequent paragraphs will outline ONLY the injection pump service work which can be accomplished without the use of special, costly pump testing equipment. If additional service work is required, the pump should be turned over to an official Diesel service station for overhaul. Inexperienced service personnel should never attempt to overhaul a Diesel injection pump.

Fig. CS115—Typical Bosch model APE injection pump.

Fig. CS116—Typical Bosch model PSB "B" or "C" injection pump.

Models 30-40-50 (Model APE)

152. TIMING TO ENGINE. Tractors equipped with a model APE pump should have the following injection timing:

Model 30 ....36° BTC
Model 40 ....34° BTC
Model 50 ....30° BTC

To check the timing of the injection pump (by overflow method) after the pump is installed as outlined in paragraph 156, proceed as follows:

Crank the engine until No. 1 piston is coming up on compression stroke and the engine flywheel mark "FPI" is in the exact center of the timing hole as shown in Fig. CS120. Using clean Diesel fuel, thoroughly wash the injection pump and injection lines. Disconnect the number one injection line from the pump. Unscrew the number one injection line connection from the pump and remove the check

Fig. CS117—Typical Bosch model PSB "Y" injection pump. Notice the external governor spring.

Fig. CS120—Models 30, 40 and 50 Diesel injection timing mark which can be viewed through the inspection port. The inspection port is on the left side on models 40 and 50.

valve spring and check valve. The spring and check valve are shown removed in Fig. CS121. Reinstall the injection line connection, leaving out the check valve and spring.

Turn the flywheel backwards ½ turn or, if more convenient, turn the flywheel forward in the direction of engine rotation 1½ turns. At this time, the fuel should run freely from the number one injection line connection when operating the hand primer of the fuel transfer pump.

Crank engine slowly in the normal direction of rotation while operating the hand primer of the fuel transfer pump, until the fuel seeps slowly from the injection line connection on top of the injection pump. Using a small tube, blow the fuel seepage away from the connection as the engine is being cranked very slowly and the hand primer being operated. Refer to Fig. CS122. By blowing the seepage away, the exact instant when the seepage stops can be observed. Cranking should be stopped the instant the fuel seepage stops. At this time, the flywheel mark "FPI" should be in the exact center of the timing hole as shown in Fig. CS120. If the mark is approaching the timing hole, the timing is advanced. If the mark has gone past the hole, the timing is retarded. To reset the timing, remove the cover from the front of the engine timing gear case and loosen the three cap screws retaining the pump drive sprocket to the pump hub. If the timing was retarded, turn the pump hub counter-clockwise (viewed from front) about one degree and recheck the timing as previously outlined. If the timing was advanced, turn the hub

Fig. CS121—Removing check valve spring (X) and valve (Y) from a Bosch model APE pump.

clockwise approximately one degree and recheck the timing. Continue this procedure until the exact timing is obtained.

When the exact timing is obtained, tighten the sprocket retaining cap screws securely and install the safety wire. Remove the number one injection line connection and reinstall the check valve and check valve spring.

153. LUBRICATION. The lower part (camshaft compartment) of the injection pump housing should be filled with a good grade of mineral base lubricating oil. The proper level can be determined by means of the oil level indicator cock.

154. TRANSFER PUMP. Model APE injection pump is equipped with a self-regulating, plunger-type transfer pump which is actuated by the injection pump camshaft. Refer to Fig. CS123.

If the pump is not operating properly, the complete pump can be renewed as a unit; or, the transfer pump can be disassembled and cleaned and checked for improved performance. Quite often, a thorough cleaning job will restore the pump to its original operating efficiency.

155. REMOVE AND REINSTALL INJECTION PUMP. Before attempting to remove the injection pump, thoroughly wash the pump and connections with clean Diesel fuel. Disconnect the injection lines from injection pump and the inlet and outlet lines from transfer pump. Disconnect the remaining lines and control rods. Cover all fuel line connections with tape or composition caps to eliminate the entrance of dirt. Crank engine until No. 1 piston is coming up on compression stroke and the flywheel mark "FPI" is in the exact center of the

Fig. CS122 — When timing a Bosch model APE injection pump by the overflow method, a small tube is used to blow the fuel seepage away while operating the hand primer (P) on the transfer pump (Q).

timing hole as shown in Fig. CS124. Remove the cover from front of the engine timing gear case and remove the three cap screws retaining the drive chain sprocket to the pump hub. Remove the pump mounting cap screws and withdraw the pump.

156. To install the injection pump, first make certain that the flywheel mark "FPI" is in the exact center of the timing hole as shown in Fig. CS124, when No. 1 piston is coming up on compression stroke. Observe the front of the pump housing and it will be noted that there are two marks (44 and 45—Fig. CS125) immediately behind the pump drive hub. One (44) is vertical and indicates top dead center of the pump; the other mark (45) is located about 1/8 turn to the right and indicates the injection pump port closing. Turn the pump hub until the port closing mark (45) is in the center of the small hole which is drilled through the pump hub. The mark can be viewed through the hole as shown at (43). With the hub so positioned, mount the pump and install the pump retaining cap screws. Place the pump drive sprocket in the drive chain so that the cap screw holes in the hub are approximately centered in the elongated holes in the sprocket and install the three cap screws, leaving out the safety wire. Turn the engine crankshaft two complete revolutions and again line up the flywheel mark "FPI" in the center of the timing hole as shown in Fig. CS124. Using an inspection mirror, check to make certain that the port closing mark (45—Fig. CS125) is centered in the pump

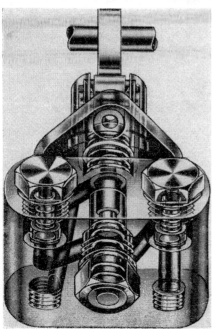

Fig. CS123—Plunger type fuel transfer pump which is used on Bosch model APE injection pumps.

Fig. CS124—Models 30, 40 and 50 Diesel injection timing mark which can be viewed through the inspection port. The inspection port is on the left side on models 40 and 50.

Fig. CS125—Bosch model APE injection pump TDC mark (44) and port closing mark (45) which can be viewed through the hole in the pump drive hub as shown at (43).

hub timing hole as shown at (43). If the mark is not centered in the hub hole, loosen the sprocket retaining screws, turn the hub until the marks are as specified and recheck the setting.

After the pump is installed, make the final timing setting by the overflow method given in paragraph 152.

157. GOVERNOR. Model APE injection pumps are equipped with an American Bosch GV, mechanical flyweight type governor. For the purposes of this manual, the governor will be considered as an integral part of the injection pump.

158. ADJUSTMENT. Recommended governed speeds are as follows:

Model 30
Max. no load engine speed...1820 rpm
Max. full load engine speed..1650 rpm
Power take-off speed at
 1650 engine rpm.......... 545 rpm
Belt pulley speed at
 1650 engine rpm ....1336 rpm
Low idle engine speed....... 625 rpm

Models 40 and 50
Max. no load engine speed...1820 rpm
Max. full load engine speed..1650 rpm
Power take-off speed
 at 1650 engine rpm........ 530 rpm
Belt pulley speed
 at 1650 engine rpm.......1000 rpm
Low idle engine speed...... 875 rpm

To adjust the governor, first start engine and run until engine is at normal operating temperature. Loosen the lock nut and remove the bumper spring adjusting screw (1—Fig. CS127,

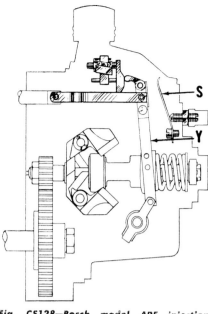

Fig. CS128—Bosch model APE injection pump governor, showing bumper spring (S) and yoke assembly (Y).

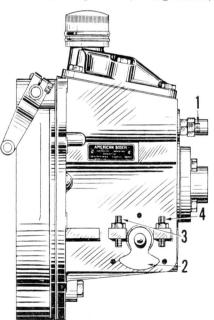

Fig. CS127—Bosch model APE injection pump, showing points for governor adjustment.
1. Bumper spring adjusting screw
2. Adjusting screw stop plate
3. High speed adjusting screw
4. Low speed adjusting screw

129 or 130). Remove the adjusting screw cover from side of injection pump. Move the operating lever stop plate (2) to the high speed position and with the stop plate contacting the high speed adjusting screw (3), turn the adjusting screw up or down to obtain the recommended high idle, no load speed. Reinstall the high idle bumper spring adjusting screw (1) and with the engine running at the high idle, no load speed, turn the adjusting screw *in* until the spring (S—Fig. CS128) just touches the fulcrum yoke assembly (Y) without an increase in engine speed. If the engine still surges at the high idle no load speed, turn the screw in slightly until surge is reduced to a minimum. Recheck the high idle, no load speed and reset, if necessary, with adjusting screw (3—Fig. CS127). Move the operating lever stop plate (2) to the low speed position and with the stop plate contacting the low idle speed

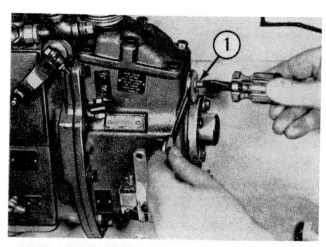

Fig. CS129—Adjusting Bosch model APE injection pump bumper spring by turning adjusting screw (1).

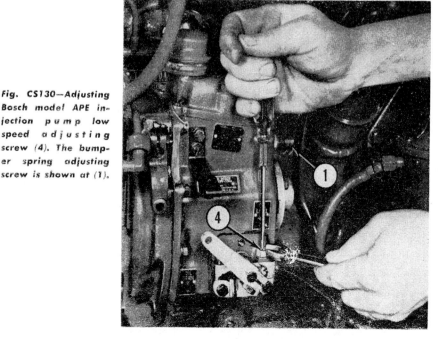

Fig. CS130—Adjusting Bosch model APE injection pump low speed adjusting screw (4). The bumper spring adjusting screw is shown at (1).

adjusting screw (4—Fig. CS127 or 130), turn the adjusting screw (4) up or down to obtain the recommended low idle speed.

Reinstall the adjusting screw cover when adjustment is complete.

Models 30-40-50 (Model PSB, Series "B", "C" or "Y")

160. TIMING TO ENGINE. Tractors equipped with a PSB series "B" or "C" injection pump should have the following injection timing.

Model 30 .... 26° BTC
Model 40 .... 28° BTC
Model 50 .... 25° BTC

All tractors equipped with a PSB series "Y" injection pump should be timed so that injection occurs at 24 degrees before top center.

Fig. CS135—Models 30, 40 and 50 Diesel injection timing mark which can be viewed through the inspection port. The inspection port is on the left side on models 40 and 50.

To check and retime the pump to the engine after the pump is installed as outlined in paragraph 165, proceed as follows:

Crank the engine until No. 1 piston is coming up on compression stroke and the engine flywheel mark "FPI" is in the exact center of the timing hole as shown in Fig. CS135. Remove plugs (P—Fig. CS137) so that the pump timing marks can be seen. If the pump timing is correct, the line mark (50—Fig. CS138) on the drive gear hub should be in register with the pointer (51) extending from the front face of the pump. If the timing marks are not in register, remove the cover from front of the engine timing gear case and loosen the three cap screws retaining the pump drive gear to the pump hub. Using a socket wrench, turn the pump hub until the timing marks are exactly in register and tighten the drive gear retaining cap screws.

162. TRANSFER PUMP. The PSB injection pumps are equipped with a positive-displacement, gear-type transfer pump which is gear driven from the injection pump camshaft. Refer to Fig. CS139. To check the operation of the transfer pump, refer to paragraph 132A.

If the pump is not operating properly, the complete pump can be renewed as a unit; or, the transfer pump can be disassembled and cleaned and checked for improved performance. Quite often, a thorough cleaning job will restore the pump to its original operating efficiency.

163. HYDRAULIC HEAD. The hydraulic head assembly (Fig. CS140) can be renewed without the use of special testing equipment. The head

Fig. CS137—To view the Bosch model PSB injection pump timing marks on models 30, 40 and 50, it is necessary to remove plugs (P).

assembly contains all of the precision components which are essential to accurate pumping, distributing, metering and delivery of the fuel. To renew the hydraulic head assembly, first wash the complete injection pump and injection lines with clean fuel oil. Remove the injection lines and disconnect the inlet and outlet lines from hydraulic head. Remove the timing window cover (20—Fig. CS 141) and crank engine until the line mark on the apex of one of the teeth on the pump plunger drive gear is in register with the "O" mark stamped on the lower face of the timing window hole as shown in Fig. CS142. Remove the two screws and carefully withdraw the control assembly being careful not to lose plunger sleeve pin (16—Fig. CS141). Remove governor cover (7) and unscrew and remove lube oil filter (8). Remove the hydraulic head retaining stud nuts and carefully with-

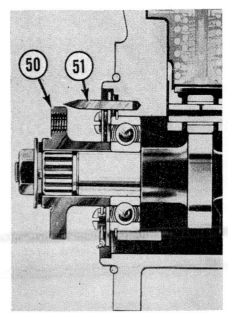

Fig. CS138—Bosch model PSB injection pump timing mark (50) is a line on edge of pump drive hub. The pointer is shown at (51).

Fig. CS139—Cut-away view of model SGB type fuel transfer pump which is used on model PSB injection pumps.

COCKSHUTT 20-30-40-50

Paragraphs 163-167

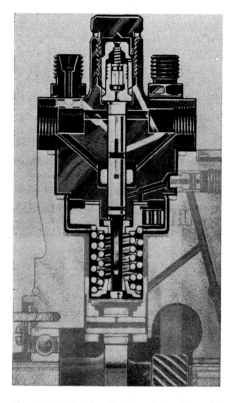

Fig. CS140—Sectional view of Bosch model PSB injection pump hydraulic head. The complete head assembly can be renewed as a unit.

draw the hydraulic head assembly from the pump housing. Do not use force when attempting to withdraw the hydraulic head. If difficulty is encountered, check to make certain that the plunger drive gear is properly positioned as shown in Fig. CS142.

When installing a new hydraulic head assembly, make certain that the line marked plunger drive gear tooth is in register with the "O" mark in the timing window as shown in Fig. CS142 and that open tooth on quill shaft gear is in register with punch mark in pump housing as shown at (A) in Fig. CS143. When installing the control sleeve assembly, the plunger sleeve pin (16—Fig. CS141) must be lined up with the slot in the control block. The remainder of the reassembly procedure is evident.

164. **REMOVE AND REINSTALL INJECTION PUMP.** Before attempting to remove the injection pump, thoroughly wash the pump and connections with clean Diesel fuel. Disconnect the injection lines from injection pump and the inlet and outlet lines from the transfer pump. Disconnect the remaining lines and control rods. Cover all fuel line connections with tape or composition caps to eliminate the entrance of dirt. Crank engine until No. 1 piston is coming up on compression stroke and the flywheel mark "FPI" is in the exact center of the timing hole as shown in Fig. CS135. Remove the pump mounting cap screws and withdraw the pump.

165. To install the injection pump, first make certain that the flywheel mark "FPI" is in the exact center of the timing hole as shown in Fig. CS135, when No. 1 piston is coming up on compression stroke. Loosen the three pump drive gear retaining cap screws. Remove the timing window cover (20—Fig. CS141) from side of pump housing. Turn the pump hub until the line mark on the apex of one of the pump plunger drive gear teeth is approximately in the center of the timing window. Then, continue turning the hub until the line mark (50—Fig. CS138) on the drive gear hub is in register with the pointer (51) extending from the front face of the pump. Mount the injection pump on the engine. The pump drive gear should be meshed with the camshaft gear so that the three drive gear retaining cap screws are approximately in the center of the elongated drive gear holes when the aforementioned timing marks are in register.

After the pump is installed, make the final timing setting as outlined in paragraph 160.

166. **GOVERNOR.** Model PSB injection pumps are equipped with a mechanical flyweight type governor. For the purposes of this manual, the governor will be considered as an integral part of the injection pump.

167. **ADJUSTMENT.** Recommended governed speeds are as follows:

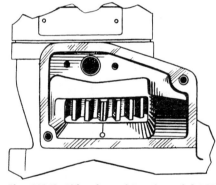

Fig. CS142—Side view of Bosch model PSB injection pump with timing window cover removed. Notice that the line on one of the gear teeth is in register with the "O" mark on the housing.

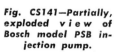

Fig. CS141—Partially, exploded view of Bosch model PSB injection pump.

2. Gasket
4. Delivery valve spring
5. Gasket
6. Hydraulic head assembly
7. Governor compartment cover
8. Filter
9. "O" ring
10. Gasket
11. Filter screen
12. Snap ring
13. Gear
14. Seal
15. Transfer pump
16. Sleeve pin
17. Snap ring
18 & 19. Control unit
20. Timing window cover
21. "O" ring
22. "O" ring
23. "O" ring
25. "O" rings

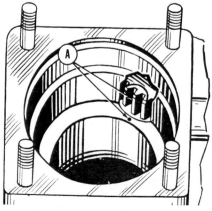

Fig. CS143—When installing a hydraulic head on a model PSB injection pump, make certain that open tooth on quill shaft gear is in register with punch mark in housing as shown at (A).

CS-51

Model 30
Max. no load engine speed...1820 rpm
Max. full load engine speed..1650 rpm
Power take-off speed at
  1650 engine rpm.........545 rpm
Belt pulley speed at
  1650 engine rpm.........1336 rpm
Low idle engine speed.....625 rpm

Models 40 and 50
Max. no load engine speed..1820 rpm
Max. full load engine speed.1650 rpm
Power take-off speed at
  1650 engine rpm.........530 rpm
Belt pulley speed at 1650
  engine rpm..............1000 rpm
Low idle engine speed.....875 rpm

To adjust the governor, first start engine and run until engine is at normal operating temperature. Then, refer to paragraph 169 for the internal governor spring series "B" and "C" pumps. Refer to paragraph 170 for the external governor spring series "Y" pumps.

169. **PSB—Series "B" and "C".** Remove the adjusting screw cover from side of injection pump. Move the operating lever stop plate (2—Fig. CS146) to the high speed position and with the stop plate contacting the high speed adjusting screw (4), turn the adjusting screw up or down to obtain the recommended high idle, no load speed.

Move the operating lever stop plate (2) to the low speed position and with the stop plate contacting the low idle speed adjusting screw (3), turn the adjusting screw up or down to obtain the recommended low idle speed.

Reinstall the adjusting screw cover when adjustment is complete.

170. **PSB—Series "Y".** The low idle speed of the engine is controlled by the position of the throttle lever and is adjusted by varying the length of a bumper spring and bolt which is located on the throttle linkage, under the fuel tank. The high idle engine speed is adjusted by nuts (N—Fig. CS147) which vary the length of the externally located governor spring.

ENERGY CELLS
Models 30-40-50

175. **R&R AND CLEAN.** The necessity for cleaning the energy cells is usually indicated by excessive exhaust smoking, or when fuel economy drops. To remove the energy cells, remove the center hood and manifold. Remove the energy cell clamp and tap the energy cell cap with a hammer to break loose any carbon deposits. Using a pair of pliers, remove the energy

Fig. CS147—The high idle speed on engines equipped with a PSB "Y" injection pump is adjusted with nuts (N).

Fig. CS150—Installing the energy cell on models 30, 40 and 50. If the surfaces (S) are rough or pitted, they can be reconditioned by lapping.

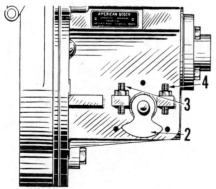

Fig. CS145—Bosch model PSB series "B" or "C" injection pump showing points for governor adjustment.
2. Adjusting screw   3. Low speed
    stop plate   adjusting screw
       4. High speed adjusting screw

cell cap. A ¼ inch tapped hole is also provided in the cap to facilitate removal.

The outer end of the energy cell body is tapped with a ⅞-14 thread to permit the use of a screw type puller when removing the cell body. The cell body can also be removed by first removing the respective nozzle. Using a brass drift inserted through the nozzle hole, bump the cell out of the cylinder head.

The removed parts can be cleaned in an approved carbon solvent. After parts are cleaned, visually inspect them for cracks and other damage. Renew any damaged parts. Inspect the seating surfaces between the cell body and the cell cap for being rough and pitted. The surfaces (S—Fig. CS150) can be reconditioned by lapping with valve grinding compound. Make certain that the energy cell seating surface in cylinder head is clean and free from carbon deposits.

When installing the energy cell, tighten the clamp nuts enough to insure an air tight seal.

PREHEATER
Models 30-40-50

176. An exploded view of the heater box is shown in Fig. CS151. Normal service includes renewing gaskets and making certain that electrical connections are tight.

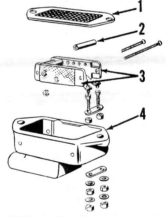

Fig. CS151—Exploded view of model 30 Diesel engine pre-heater unit. Models 40 and 50 are similar.
1. Screen   3. Heater element
2. Element spacer   4. Heater box

NON-DIESEL GOVERNOR
ADJUSTMENT
Model 20

Non-Diesel governors are of the centrifugal flyball (flyweight on 30 models prior to serial No. 9953) type and are gear driven from the camshaft gear. On model 20, non-Diesel model 30 and some early model 40 tractors, the hydraulic pump is mounted on and driven by the governor.

180. Before attempting any governor adjustments, first remove any binding or lost motion from the operating linkage. With engine stopped, remove the clevis pin which attaches the governor-to-carburetor link rod to

the governor lever. Move the carburetor throttle lever to a point where the throttle butterfly is in the wide open position and adjust the length of the carburetor-to-governor link rod so that clevis pin will freely slide into position. To adjust the governed speed, first start engine and bring to normal operating temperature. Move the speed control hand lever (1—Fig. CS160) fully forward and adjust the limiting screw (2) to provide the desired high idle speed.

Hunting (surging) or unsteady running at no load speeds can be eliminated by turning the bumper spring adjusting screw (3) *in* until the undesirable condition is eliminated.

CAUTION: The bumper spring adjusting screw should never be turned in far enough to raise the engine speed. Note: If unsteady running cannot be eliminated by this procedure, check for malfunctioning governor, binding linkage and/or faulty carburetor adjustments.

Recommended operating speeds are as follows:

Max. no load engine speed ..2000 rpm
Max. full load engine speed ..1800 rpm
Power take-off speed at
 1800 engine rpm .... 563 rpm
Belt pulley speed at
 1800 engine rpm .... 1160 rpm
Low idle engine speed
 (gasoline) .... 475 rpm
Low idle engine speed
 (distillate) .... 575 rpm

Models 30-40-50

181. Before attempting any governor adjustments, first remove any binding or lost motion from the operating linkage. An adjustable ball and socket joint is fitted to each end of the governor-to-carburetor link rod. These joints should be well lubricated and should be adjusted to provide a snug fit without causing friction. Disconnect the link rod from the governor arm and check the overall length of the link rod as indicated by dimension (A—Fig. CS161). Adjust the length of the rod until dimension (A) is 11 7/16 inches for model 30, 16 inches for models 40 and 50. To obtain the desired 16 inches overall length on model 40 tractors prior to serial No. 3224, it may be necessary to install a new, longer link rod or lengthen the present one by cutting and welding in an extra piece of rod.

Note: On model 40 tractors prior to serial No. 6219, the carburetor throttle lever was positioned on the throttle shaft by a clamp screw. On such models, it is important to check the relative position of the lever with respect to the shaft as follows: With the throttle butterfly completely closed, the center line of the throttle levers should be tilted 12 degrees rearward from the center line of the carburetor mounting flange as shown in Fig. CS162. If the adjustment is incorrect, loosen the clamp screw and re-position the lever.

With one end of the link rod connected to the carburetor throttle lever and with the governor spring and spacer removed, move the governor arm fully forward (toward radiator) and pull the throttle link rod to its extreme forward position. With the parts so positioned, the link screw should drop straight into the hole in the governor arm. If the screw does not drop straight in, but is within 1/16 inch either way, adjust the length of the rod until proper register is obtained.

Note: If the length of the rod at 11 7/16 inches for model 30 or 16 inches for models 40 and 50 is not within 1/16 inch either way of making the connection, DO NOT adjust the linkage rod. The governor arm should be bent either way as required to make the connection.

On all models, reinstall the governor spring and spacer. On model 40 tractors so equipped and all model 50 trac-

Fig. CS160—Right side view of model 20 tractor, showing the speed control hand lever (1), speed limiting screw (2) and the bumper spring adjusting screw (3).

Fig. CS161—Governor-to-carburetor link rod on model 30 tractors. The rod is similar on models 40 and 50. The overall length of the rod (dimension A) should be 11 7/16 inches on model 30, and 16 inches on models 40 and 50.

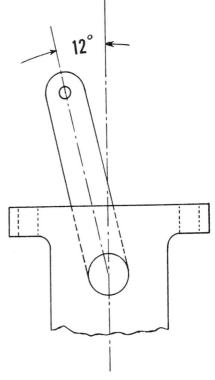

Fig. CS162—On model 40 tractors prior to Ser. No. 6219, check the position of the carburetor throttle lever to make certain that the lever is tilted 12 degrees rearward from carburetor center line when throttle butterfly is closed.

tors, remove bumper spring from the carburetor rear mounting bolt. On all models, start engine and bring to operating temperature. Adjust the speed regulating screw to provide the recommended high idle speed. On early model 30 tractors, the regulating screw is located in a bracket which fastens to the governor housing. On late model 30 tractors, the regulating screw is located at the top of an adapter housing which is directly behind the governor. On models 40 and 50, the speed is adjusted by increasing or decreasing the governor spring tension by means of the spring tension adjusting screw.

On models 40 and 50, reinstall the bumper spring on the carburetor rear mounting bolt and adjust the spring so that it rests lightly on the throttle lever when engine is running at 1850 rpm. On model 30 tractors, adjust the length of the eye screw to which the governor spring is hooked, to eliminate surging or unsteady running.

Note: If unsteady running cannot be eliminated by adjusting the bumper spring on models 40 and 50 or by adjusting the length of the eye screw on model 30, check for malfunctioning governor, binding linkage and/or faulty carburetor adjustments.

Recommended operating speeds are as follows:

Model 30
Max. no load engine speed...1810 rpm
Max. full load engine speed..1650 rpm
Power take-off speed at
 1650 engine rpm .... 545 rpm
Belt pulley speed at
 1650 engine rpm ....1336 rpm
Low idle engine speed
 (gasoline & LP-gas)...... 425 rpm
Low idle engine speed
 (distillate) .... 525 rpm

Models 40 and 50
Max. no load engine speed..1850 rpm
Max. full load engine speed..1650 rpm
Power take-off speed at
 1650 engine rpm .... 530 rpm
Belt pulley speed at
 1650 engine rpm ....1000 rpm
Low idle engine speed
 (gasoline & LP-gas)...... 425 rpm
Low idle engine speed
 (distillate) .... 525 rpm

REMOVE AND REINSTALL
Model 20

190. To remove the governor, first drain cooling system and remove front and center hood as a unit. Remove radiator, disconnect governor linkage and remove governor.

After governor is installed adjust the unit as outlined in paragraph 180.

Models 30-40-50

191. To remove the governor, first drain cooling system and disconnect radiator hose from inlet elbow on left side of cylinder head. Disconnect governor linkage and remove governor.

Note: The governor shaft is carried in a renewable split type bushing which is located in an adapter housing bolted to the rear face of the engine front end plate. On model 30 non-Diesel tractors equipped with a hydraulic system, it is necessary to remove the hydraulic pump before the adapter housing can be removed.

After installation, ream the bushing to an inside diameter of 1.500-1.501. This will provide a clearance of 0.0015-0.0035 for the 1.4975-1.4985 diameter governor shaft.

Before installing the governor, remove cover (or hydraulic pump on model 30 non-Diesels) from rear face of the adapter housing. Install governor and tighten the cap screws. With one hand, move the governor lever back and forth; while pushing on rear of governor shaft with the other hand as shown in Fig. CS166. This procedure will determine whether the governor shaft has a free fit in the adapter housing bushing. If the shaft does not have a slight amount of fore and aft movement in the bushing, loosen the adapter housing bolts and using a raw hide hammer, tap the housing either way as required until the governor shaft is free in the bushing.

Adjust the governor as outlined in paragraph 181.

OVERHAUL
Models 20-30-40-50

192. To overhaul the governor, first remove the unit as outlined in paragraphs 190 or 191. The procedure for disassembling and reassembling the governor is evident after an examination of the unit and reference to Figs. CS165, 167 or 168. After governor is disassembled, thoroughly clean all

Fig. CS166—Checking freedom of governor shaft in the adapter housing bushing on models 30, 40 and 50. A binding condition is eliminated by shifting the adapter housing. See text.

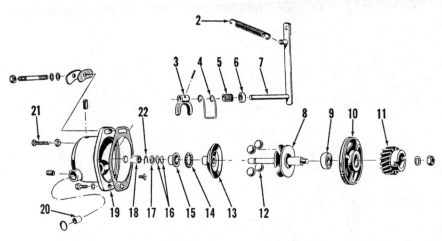

Fig. CS165—Exploded view of model 20 governor. The hydraulic pump, on models so equipped, is driven from the governor shaft.

2. Governor spring
3. Fork
4. Bumper spring
5. Needle bearing
6. Oil seal
7. Governor lever
8. Drive shaft
9. Ball bearing
10. Governor base
11. Drive gear
12. Flyballs
13. Cupped race
14. Thrust bearing
15. Fork base
16. Shims
17. Stop washer
18. Bushing
19. Housing
20. Bushing
21. Bumper spring adjusting screw
22. Snap ring

COCKSHUTT 20-30-40-50

parts and renew any which are excessively worn. On flyweight type governors, shown in Fig. CS167, inspect weights, weight pins and thrust sleeve for wear. On flyball type governors shown in Figs CS165 or 168, make certain that the ball driver is tight on the shaft and that the ball race is free on the shaft. When assembling flyball governors, there should be a space of 0.004-0.006 between the ball driver and the flat race and, when balls are "in", the cupped race (6—Fig. CS168) or (13—Fig. CS165) should have a fore and aft movement on the shaft of 0.230-0.240. Obtain this latter dimension by varying the number of 0.010 shims (11—Fig. CS168 or 16—Fig. CS165). For replacement purposes, the governor shaft, gear, spider and flat race are available as an assembled unit.

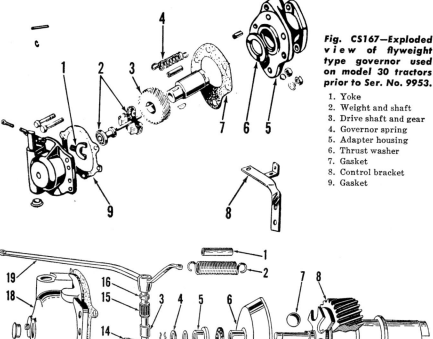

Fig. CS167—Exploded view of flyweight type governor used on model 30 tractors prior to Ser. No. 9953.

1. Yoke
2. Weight and shaft
3. Drive shaft and gear
4. Governor spring
5. Adapter housing
6. Thrust washer
7. Gasket
8. Control bracket
9. Gasket

Fig. CS168—Exploded view of flyball type governor used on models 30, 40 and 50.

1. Spring spacer
2. Governor spring
3. Spacer
4. Stop washer
5. Fork base
6. Cupped race
7. Flyballs
8. Drive shaft and gear
9. Thrust washer
10. Thrust bearing
11. Shims
13. Needle bearing
14. Fork
15. Needle bearing
16. Oil seal
18. Housing
19. Governor arm

When installing the radiator, make certain that shock pads are installed under the radiator mounts.

THERMOSTAT
Models 20-30-40-50

196. The thermostat on model 20 tractors can be removed after removing the water outlet casting from top of cylinder head. The thermostat can be removed from model 30, 40 and 50 tractors after removing the thermostat housing from front of the water outlet casting which is located on top of cylinder head.

WATER PUMP
Model 20

197. **RESEAL.** The water pump seal assembly (3—Fig. CS170) can be renewed without removing the complete pump unit from the tractor. Remove front and center hood and radiator and proceed as follows: Remove fan blades and fan belt. Remove nuts retaining drive support (6) to pump body and withdraw the support assembly. Remove set screw (4) retaining impeller to pump shaft and using a suitable puller, remove the impeller. The seal assembly which is contained in the front part of the impeller can be renewed at this time. Inspect the seal contacting surface of bushing (5). If the surface is pitted or rough, renew the bushing.

198. **OVERHAUL.** To overhaul the water pump, first remove the impeller and seal assembly as outlined in paragraph 197. Remove the nut retaining pulley to the drive shaft and using a suitable puller or press, remove the pulley. Loosen lock nut (15—Fig. CS170), remove set screw (14) and press the shaft and bearings assembly forward and out of pump body. The need and procedure for further disassembly is evident after an examination of the unit.

COOLING SYSTEM

RADIATOR
Models 20-30-40-50

195. The radiator can be removed from models 20 and 30 after draining the cooling system and removing the front and center hood as a unit.

To remove the radiator on models 40 and 50, first drain cooling system and remove front and center hood as a unit. Remove steering wheel. Remove adjusting cap from front of steering gear housing, extract cotter pin from steering shaft and turn the worm and steering shaft forward and out of the gear housing. The radiator can be removed at this time.

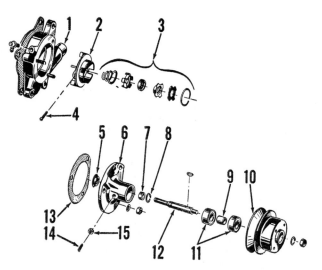

Fig. CS170—Exploded view of model 20 water pump. The pump can be overhauled without removing body (1) from tractor.

1. Pump body
2. Impeller
3. Seal assembly
4. Set screw
5. Bushing
6. Drive support
7. Bearing retainer
8. Retainer ring
9. Spacer
10. Fan hub and pulley
11. Bearings
12. Shaft
13. Gasket
14. Set screw
15. Lock nut

When installing the ball bearings (11), make certain that open (unshielded) side of bearings face each other and pack the bearings with a good quality fibrous type grease.

Models 30-40-50

199. RESEAL AND OVERHAUL. The pump cannot be overhauled without removing the pump from the tractor. To remove the pump, first remove the radiator. Note: On some model 30 tractors, it is not absolutely necessary to remove the radiator; however, time will usually be saved by doing so. Remove fan blades and pump pulley and unbolt pump from cylinder head. Using a press or suitable puller, remove pulley hub as shown in Fig. CS172. Remove cover from back of pump and using a pair of long nose pliers, remove the pump shaft retaining snap ring. Press the drive shaft and bearing unit out of impeller and pump body as shown in Fig. CS173.

Renew bearing and shaft unit (2—Fig. CS174) if shaft is bent or if bearing is a loose fit in pump body. Inspect seal contacting surface in pump body. If the surface is pitted or rough, either recondition the surface or renew the pump body. Always renew a questionable seal.

When reassembling, press the shaft and bearing unit into the pump body and install the retaining snap ring. Assemble seal into impeller and start the impeller on the shaft. Place pump in a vise as shown in Fig. CS175 and press the impeller on the drive shaft until end of shaft is flush with rear surface of the impeller. Press pulley hub on front of drive shaft and install cover.

Note: A newly overhauled pump might leak for the first few hours of operation. This condition should correct itself after the parts have had an opportunity to wear in.

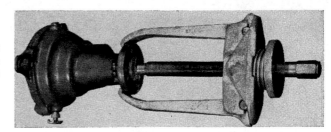

Fig. CS172—Removing pulley hub from models 30, 40 and 50 water pump.

Fig. CS173 — Pressing models 30, 40 and 50 water pump drive shaft and bearing unit out of impeller and housing.

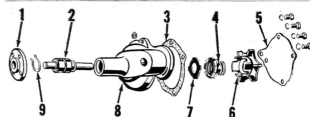

1. Pulley hub
2. Drive shaft
3. Gasket
4. Seal
5. Cover
6. Impeller
7. Washer
8. Pump body
9. Snap ring

Fig. CS174—Exploded view of models 30, 40 and 50 water pump. The pump can be removed after removing the radiator.

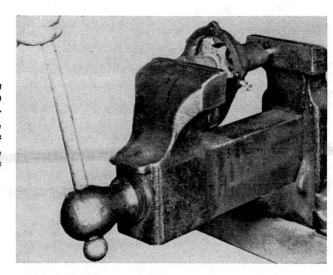

Fig. CS175 — Pressing models 30, 40 and 50 water pump impeller on the pump drive shaft. Rear face of impeller should be flush with end of shaft.

IGNITION AND ELECTRICAL SYSTEM

DISTRIBUTOR
Models 20-30-40-50

200. APPLICATIONS. Auto-Lite ignition distributors are used and their applications are as follows:

Model 20 .... IAD 6004-1C
Model 30 .... IAD 6003-1A
Model 40 .... IAD 6001-1B
Model 50 .... IAD 6001-1B

Refer to Distributor Section of STANDARD UNITS for testing and general overhaul data.

201. INSTALLATION AND TIMING. The battery ignition distributor is timed in the full retard position on all models. To install and time the distributor, first adjust breaker contact gap to 0.020; then, crank engine until No. 1 piston is coming up on compression stroke and continue cranking until the flywheel timing marks are in position.

Model 20 Gasoline. The static spark should occur 6 degrees before top dead center. The flywheel is in the proper position when the sixth degree graduation before the "DC" mark is in the exact center of the timing hole on right side of engine as shown in Fig. CS180.

Model 20 Distillate. The static spark should occur when "DC" mark on flywheel is in the exact center of the timing hole on right side of engine as shown in Fig. CS181.

Model 30 Gasoline, Distillate and L.P. Gas. The static spark should occur 10 degrees before top dead center. The flywheel is in the proper position when the "STATIC SPARK" mark on flywheel is in the exact center of the timing hole on right side of engine as shown in Fig. CS183.

Models 40 and 50. The static spark should occur when "T.D.C." mark on flywheel is in the exact center of the timing hole on left side of engine.

Turn the distributor drive shaft until rotor arm is in the number one firing position and mount the ignition unit on the engine. On models 30, 40 and 50, make certain that the distributor timing pointer is on the zero degree graduation. Remove ignition cable from No. 1 spark plug and hold free end of cable near engine block. Loosen distributor clamp and, with ignition switch turned on, turn distributor body slowly until a spark occurs at end of spark plug cable; then lock the distributor in this position.

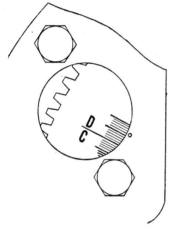

Fig. CS180—Position of ignition timing marks for model 20 gasoline. The static spark should occur 6 degrees before top center as shown.

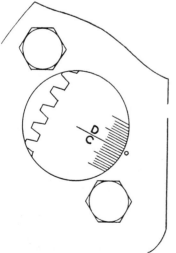

Fig. CS181—Position of ignition timing marks for model 20 distillate. The static spark should occur at top center.

The efficiency and the amount of power that the engine has at rated rpm depends on the proper operation of the automatic spark advance unit which is built into the distributor. If the engine lacks power at rated rpm, check the distributor advance curve as given in the Distributor Section of STANDARD UNITS.

GENERATOR & REGULATOR
Models 20-30-40-50

202. Auto-Lite generators are used on all models and their applications are as follows:

Model 20 .... GHD-6001K
Model 30 (Non-Diesel) prior Ser. No. 31196 .... GEO-4818A3
Model 30 (Non-Diesel) after Ser. No. 31195 .... GHD-6001F
Model 30 (Diesel) prior Ser. No. 34007 .... GEF-4813B
Model 30 (Diesel) after Ser. No. 34006 .... GHH-6001B
Model 40 (Non-Diesel) prior Ser. No. 6786 .... GEO-4818A3
Model 40 (Non-Diesel) after Ser. No. 6785 .... GHD-6001F
Model 40 (Diesel) prior Ser. No. 8666 .... GEF-4813B
Model 40 (Diesel) after Ser. No. 8665 .... GHH-6001B
Model 50 (Non-Diesel) .. GHD-6001F
Model 50 (Diesel) GHH-6001B

The following generators may be found on late production tractors:
Model 20 .... GHD-6001Q
Models 30, 40 & 50 (Non-Diesels) GHD-6001Y
Models 30, 40 & 50 (Diesels) GHH-6001E

Refer to Generator Section of STANDARD UNITS for testing data.

203. Auto-Lite regulators are used and their applications are as follows:

Fig. CS183—Position of ignition timing mark "SPARK" for model 30. The static spark should occur 10 degrees before top center.

Model 20 .... VRR-4102B
Model 30 (Non-Diesel) prior Ser. No. 31196 .... CB-4014
Model 30 (Non-Diesel) after Ser. No. 31195 .... VRR-4102A
Model 30 (Diesel) prior Ser. No. 34007 .... VRS-5201AX
Model 30 (Diesel) after Ser. No. 34006 .... VRT-4102A
Model 40 (Non-Diesel) prior Ser. No. 6786 .... CB-4014
Model 40 (Non-Diesel) after Ser. No. 6785 .... VRR-4102A
Model 40 (Diesel) prior Ser. No. 8666 .... VRS-5201AX
Model 40 (Diesel) after Ser. No. 8665 .... VRT-4102A
Model 50 (Non-Diesel) .. VRR-4102A
Model 50 (Diesel) VRT-4102A

The following regulators may be found on late production tractors:
Model 20 and Non-Diesel 30, 40 & 50 models VRR-4103A
Models 30, 40 & 50 (Diesels) VRT-4103A

Refer to Regulator Section of STANDARD UNITS for testing data.

STARTING MOTOR
Models 20-30-40-50

204. Auto-Lite starting motors are used and their applications are as follows:

Model 20 .... MZ-4072
Model 30 (Non-Diesels) .. MAX-4081
Model 30 (Diesels) MBR-4022
Models 40 & 50 (Non-Diesels) MAX-4078
Models 40 & 50 (Diesels) .. MBR-4023

Refer to Starting Motor Section of STANDARD UNITS for testing data.

ENGINE CLUTCH
APPLICATIONS
Models 20-30-40-50

210. All models are equipped with Borg & Beck single plate, spring loaded, dry type clutches as follows:

Tractor Model	Clutch Model	Borg-Warner Cover Assy. No.
20	9A7	972
30	9A7	981
40	11A6	879
50	11A6	879

Refer to Clutch Section of STANDARD UNITS for overhaul and release lever setting data.

Model 30 tractors are available with a continuous (live) power take-off as optional equipment. When the tractors are so equipped, a splined spider (A—

Fig. CS200) and ring assembly is bolted to the clutch cover plate as shown.

ADJUSTMENT
Models 20-30-40-50

211. Adjustment to compensate for lining wear is accomplished by adjusting the clutch pedal linkage, NOT by adjusting the position of the clutch release levers. The clutch is properly adjusted when the clutch pedal has a free travel of 1-1⅛ inches.

On model 20, make this adjustment by varying the length of the clutch release rod (6—Fig. CS201) by means of clevis (7) at rear end of rod.

On models 30, 40 and 50, make this adjustment by varying the length of the clutch release rod (X — Fig. CS202) by means of clevis (Y) at front end of rod.

REMOVE AND REINSTALL CLUTCH
Model 20

212. The removal procedure which will be subsequently outlined includes removal of the tractor deck plate. An alternate procedure which is used by some mechanics, does not include removal of the deck plate and the clutch is removed from below. It is found, however, that removing the deck plate will be more convenient and will usually save some time.

To remove the clutch, proceed as follows: Remove battery cover, disconnect battery cables and remove battery. Remove deck plate (sheet metal covering over clutch shaft) from tractor. Remove the clevis pin from rear end of clutch release rod and remove the release rod (6—Fig. CS201). Disconnect the clutch shaft coupling and slide the clutch shaft (10) rearward. Unbolt clutch housing cover (5) and remove cover and clutch shaft as a unit.

Note: The clutch release bearing can be renewed at this time.

Unbolt the clutch cover plate (1) from flywheel and remove the cover assembly and lined plate. Notice that the clutch pilot shaft is an integral part of the lined plate. If the pilot shaft has an excessive amount of clearance in the pilot bushing, renew the bushing.

Install the clutch by reversing the removal procedure and adjust the unit as outlined in paragraph 211.

Model 30

213. There are two procedures for removing the clutch on model 30 tractors. The first and probably the most used method involves splitting the engine frame from the intermediate case as outlined in the subsequent paragraph 214; then, removing clutch from flywheel. The second method involves moving the engine forward in the engine frame enough to clear the clutch shaft. To move the engine forward, perform the work outlined in R&R ENGINE, paragraph 41.

214. **SPLIT ENGINE FRAME FROM INTERMEDIATE CASE.** To split engine frame from the intermediate case, first drain cooling system and remove center hood; or, if more convenient, remove front and center hood as a unit. On distillate models, disconnect the radiator shutter control wire at shutters. On models so equipped, disconnect hour meter wires from the oil pressure switch. On non-Diesel models, drain hydraulic system, disconnect lines from hydraulic pump, remove battery box cover, disconnect battery cables from battery and remove battery and battery box. On all

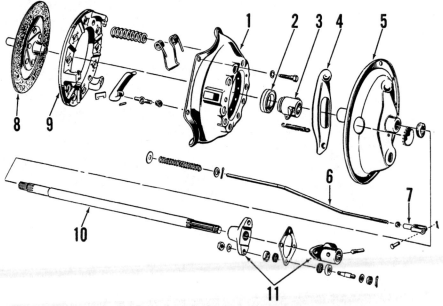

Fig. CS201—Exploded view of model 20 clutch, clutch shaft and associated parts.
1. Clutch cover plate
2. Release bearing
3. Bearing carrier
4. Release arm
5. Housing cover
6. Release rod
7. Clevis
8. Lined plate
9. Pressure plate
10. Clutch shaft
11. Coupling

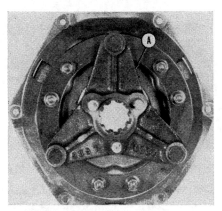

Fig. CS200—Rear view of model 30 clutch as used on models with continuous (live) power take-off. The PTO drive spider is shown at (A).

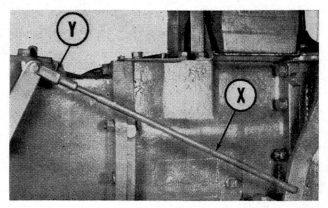

Fig. CS202—Models 40 and 50 clutch rod (X) and adjusting clevis (Y). Model 30 is similarly constructed. Adjust the length of the clutch rod to provide a clutch pedal free travel of 1-1⅛ inches.

models, disconnect wire from voltage regulator, cable from starting motor and heat indicator bulb from water outlet casting on top of cylinder head. On gasoline and distillate models, shut off fuel and disconnect fuel lines and choke rod from carburetor. On all non-Diesel models, disconnect rod from starting motor, disconnect governor control rod at governor, disconnect distributor wire from coil and remove the carburetor air intake tube. On all models, disconnect oil gage line at junction block on left side of cylinder head, unbolt tail light wire clips and disconnect tail light wire at tail light. On L.P. gas models, disconnect and remove water pipes, vapor hose and balance line from converter. On Diesel models, remove the manifold air intake tube, disconnect heater cable from manifold and disconnect speed control rod (or rods) and the fuel supply and return lines. Using a punch and hammer, remove the pin retaining the rear universal joint to the steering shaft and bump the universal joint free from the shaft. Disconnect fuel tank brackets from tractor and lift fuel tank, steering wheel and shaft, instrument panel, etc., as a unit from tractor. The removed unit on gasoline and distillate models is shown in Fig. CS203. Remove clutch housing cover and starting motor. Disconnect clutch release rod. Support tractor under engine frame and transmission case and remove cap screws and dowel bolts retaining engine frame to the intermediate case. Using a suitable pry bar, split the tractor.

215. After splitting the tractor, or moving the engine forward, the clutch release bearing can be renewed.

Remove the cap screws retaining the clutch cover plate to the flywheel and remove the cover assembly and lined plate. The clutch shaft pilot bearing can be renewed at this time.

When reinstalling the clutch, use a spare clutch shaft to align the driven plate splines with respect to the clutch shaft pilot bearing and adjust the clutch pedal linkage as outlined in paragraph 211.

Models 40-50

216. To remove the clutch, it is first necessary to split engine frame from transmission case as outlined in the subsequent paragraph 217.

217. **SPLIT ENGINE FRAME FROM TRANSMISSION CASE.** To split the engine frame from the transmission case, first support both halves of tractor separately and disconnect the clutch release rod and the tail light wire. Remove belt pulley assembly, or remove the pulley hole cover from top of transmission case. Remove the clutch housing cover and disconnect battery cable from battery. Disconnect the fuel tank rear bracket from the transmission case. Remove bolts and cap screws attaching engine frame to transmission case and using a suitable pry bar, split the tractor.

218. After splitting the tractor as in the preceding paragraph, the clutch release bearing can be renewed.

Remove the cap screws retaining the clutch cover plate to the flywheel and remove the cover assembly and lined plate. Model 40 tractors prior to serial No. 11238 and model 50 tractors prior to serial No. 1297 were equipped with a pilot bushing (28—Fig. CS205) for the front end of the clutch shaft. This bushing can be renewed at this time. On later models, a ball type pilot bearing is used and it can be renewed at this time. The early model bushing type can be, and should be, converted to the late model ball type by proceeding as follows: Remove the three cap screws retaining the input shaft sleeve (34) to front of transmission case and remove sleeve and clutch shaft. Rework front end of clutch shaft (24) in accordance with the dimensions shown in Fig. CS206. Then, install the new pilot bearing and flywheel coupling kit which is available under Cockshutt part No. TO-14014.

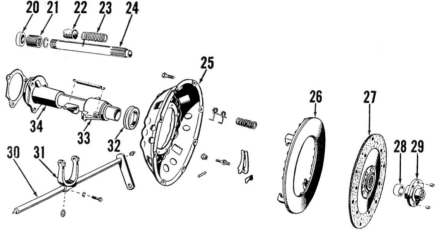

Fig. CS205—Exploded view of models 40 and 50 clutch, clutch shaft and associated parts. Items (22) and (23) are not used when tractor is equipped with power take-off.

20. Spacer	24. Clutch shaft	28. Pilot bushing (or bearing)
21. Collar	25. Clutch cover plate	29. PTO coupling
22. Felt oil seal	26. Pressure plate	30. Release shaft
23. Spring for seal	27. Lined disc	32. Release bearing
		33. Bearing carrier
		34. Bearing cap and sleeve

Fig. CS203—Model 30 instrument panel, fuel tank, etc., removed from tractor as a unit.

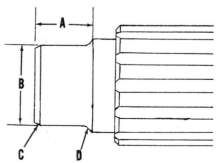

Fig. CS206—Models 40 & 50 clutch shaft reworking dimensions. This is necessary when converting from a bushing to a ball type pilot bearing.

A. ¾ inch
B. 1.3765-1.3775
C. 3/64 inch by 30° chamfer
D. 1/16 inch radius

When reinstalling the clutch, use a spare clutch shaft to align the driven plate splines with respect to the clutch shaft pilot bearing and adjust the clutch pedal linkage as outlined in paragraph 211.

CLUTCH RELEASE BEARING

Models 20-30-40-50

219. On all models, the clutch release bearing can be renewed by following the procedure for R&R of clutch. For model 20, refer to paragraph 212, for model 30, refer to paragraphs 213, 214 and 215 and for models 40 and 50, refer to paragraphs 216, 217 and 218.

CLUTCH SHAFT

Model 20

220. On model 20 tractors, the clutch shaft is removed when removing the clutch. Refer to paragraph 212.

Model 30

221. To remove the clutch shaft, first split the engine frame from the intermediate case as outlined in paragraph 214. Then, remove the release bearing and the release bearing carrier. Remove the clutch shaft cap from front of intermediate case and on models equipped with continuous (live) power take-off, remove bearing and power take-off drive sleeve and gear from front of the intermediate case. On models so equipped, remove the belt pulley assembly, and on Diesel models equipped with a hydraulic system, remove the hydraulic pump. On all models, remove the top cover from the intermediate case. On models prior to Serial No. 25461, work through the pulley compartment opening and remove the safety-wired cap screw and lock plate (23—Fig. CS207) which retains the clutch shaft rear bearing in the middle wall in the intermediate case. On models after Serial No. 25460, the clutch shaft rear bearing is retained by bolt (24); and the castellated nut for same is removed through the creeper gear compartment opening. On models so equipped, remove the safety wired set screw retaining the creeper gear shifter fork to the shifter shaft and withdraw the shifter shaft. If a new rear snap ring (26) is available, use a cold chisel and split the snap ring. If a new snap ring is not available, release the snap ring from the groove in the clutch shaft, and while pulling the clutch shaft forward, work the snap ring off the rear end of the clutch shaft. On models not equipped with continuous (live) power take-off, the clutch shaft can be withdrawn from front of intermediate case at this time. On models that are equipped with live power take-off, pull clutch shaft forward until the bevel gear (22) contacts the front drive gear on the live power take-off shaft. Move front of clutch shaft up and toward left side of tractor. Turn shaft enough for bevel gear (22) to clear the live power take-off gear and withdraw the clutch shaft. On models prior to serial No. 3151, check the pilot bearing in the forward end of the transmission main drive shaft. On later models, the pilot bearing (31) is located in the rear end of the clutch shaft.

To install the clutch shaft, reverse the removal procedure. The mesh position of the spiral bevel pinion (22) is non-adjustable with respect to its mating spiral bevel gear. The backlash, however, should be adjusted as outlined in paragraph 325A.

Models 40-50

222. To renew the clutch shaft, first split the engine frame from the transmission case as outlined in paragraph 217. Unbolt the transmission input shaft front bearing cap and sleeve (34—Fig. CS205) from the front of the transmission case and withdraw the sleeve and clutch shaft (24). The forward end of the clutch shaft on model 40 tractors prior to serial No. 11238 and model 50 tractors prior to serial No. 1297 is piloted in a bushing (28) which is retained in the flywheel coupling. On later models, the shaft is piloted in a ball bearing. Refer to paragraph 218 for information concerning the installation of the ball type pilot bearing on earlier models.

TRANSMISSION

Although most transmission repair jobs require overhauling the complete unit, there are infrequent instances where the failed or worn part is so located that repair can be accomplished without complete disassembly of the transmission. When such cases occur, considerable time will be saved by following the procedure given under the heading of "Basic Procedures" for the respective tractors.

Model 20

The transmission, differential, main drive bevel pinion and ring gear (crown gear) are all contained in the same case to which are attached the final drive and differential shaft (bull pinion shaft) housing assemblies. A wall in the common case separates the differential compartment from the transmission compartment. The combination belt pulley and power take-off unit, when used, is mounted on the rear of the transmission case and the unit is driven by the transmission main shaft (26—Fig. CS211).

230. **REMOVE AND REINSTALL TRANSMISSION ASSEMBLY.** To remove the transmission from the tractor, first support rear of tractor under frame rails and block front end so that tractor has no tendency to tip. Install a rolling stand under transmission case. Remove both rear wheel and tire units, disconnect brake rods and tail light wires and remove fenders. Unbolt and remove both final drive housings from the transmission case. Remove the transmission case rear cover, or if tractor is so equipped, remove the combination belt pulley and power take-off unit. Remove battery cover, disconnect battery cables and remove battery. Remove deck plate (sheet metal covering over clutch shaft) from tractor. Disconnect the clutch shaft coupling. Remove seat, disconnect hydraulic lines and remove the hydraulic lift unit from top of transmission case. Unbolt the frame side rails from transmission case and frame brackets (fish plates). Roll transmission rearward and away from tractor.

The transmission can be installed by reversing the removal procedure.

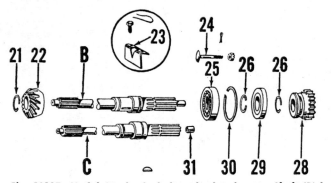

Fig. CS207—Model 30 clutch shaft and related parts. Shaft (B) is used on tractors prior to Ser. No. 3151. Shaft (C) is used on later models.

21. Snap ring
22. Belt pulley drive pinion
23. Bearing retainer (used prior to Ser. No. 25461)
24. Bearing retainer bolt (used after Ser. No. 25460)
25. Bearing
26. Snap ring
28. Creeper sliding gear
29. Oil seal
30. Snap ring
31. Pilot bearing

231. **BASIC PROCEDURES.** Preliminary data for removing the various transmission components are given in the following paragraphs. Refer to Fig. CS211 for a sectional view of the transmission.

232. *Shifter Rails and Forks.* The transmission shifter rails and forks which are attached to the transmission side cover can be removed after removing the cover.

To remove the cover, support tractor under frame rails and block front end so that tractor has no tendency to tip. Install a rolling floor jack under front of transmission case. Remove battery cover, disconnect battery cables and remove battery. Remove deck plate (sheet metal covering over clutch shaft) from tractor. Disconnect the clutch shaft coupling, brake rods and tail light wires. Unbolt transmission case and frame brackets (fish plate) from side rails and roll transmission assembly rearward. Remove the right hand wheel and tire unit and fender. Unbolt and remove the right final drive assembly from transmission case. Remove the right hand frame bracket (fish plate) from side of transmission case. The side cover and shifter rails and forks can be removed after withdrawing the shifter lever.

233. *Reverse Idler.* The reverse idler gear and shaft can be removed after removing the transmission case side cover as in the preceding paragraph and the transmission case rear cover or combination belt pulley and power take-off unit.

234. *Bevel Pinion Shaft.* The main drive bevel pinion shaft (35—Fig. CS211) can be removed after removing the following component units. Remove transmission as outlined in paragraph 230. Remove the transmission case side cover. Remove the differential bearing carriers and withdraw the differential and main drive bevel ring gear assembly.

235. *Main Shaft.* The transmission main shaft (26—Fig. CS211) can be removed after removing the reverse idler gear and shaft.

236. **MAJOR OVERHAUL.** A detailed procedure for removing and overhauling the various transmission components is given in the following paragraphs. Before attempting to overhaul the transmission, perform the following preliminary work. Remove the transmission as outlined in paragraph 230. Remove both differential bearing carriers, being careful not to mix or lose shims which are installed between the differential bearing carriers and the transmission case. Remove the differential and main drive bevel ring (crown) gear assembly. Remove the right hand frame bracket (fish plate) from side of transmission case.

237. **SHIFTER RAILS AND FORKS.** After performing the work in paragraph 236, raise the shifter lever dust boot and using a thin blade screw driver, release snap ring (2—Fig. CS212). Pull shifter lever assembly up and out of case. Remove cover and shifter rails and forks assembly from side of transmission case.

The procedure for disassembling and reassembling the shifter rails and forks is evident after an examination of the unit and reference to Fig. CS212. Be careful, however, not to lose the balls (13) and springs (12) as shifter rails are removed. After the unit is disassembled, check all parts and renew any which are excessively worn.

When installing the cover, place the transmission gears in neutral position and make certain that shifter lever gates in the forks are aligned.

238. **REVERSE IDLER.** The reverse idler gear and shaft can be removed after removing the transmission case side cover. Working through the rear opening in the transmission case, remove the cap screw and lock plate (34—Fig. CS213) which retains the reverse idler shaft (33) in the transmission case. Withdraw the shaft from the rear of case and remove the gear through the side cover opening. The reverse idler gear bushing (32) can be renewed at this time. Ream the bushing after installation to an inside diameter of 0.740-0.741. The 0.7375-0.738 diameter reverse idler gear shaft should have a clearance of 0.002-0.0035 in the bushing.

239. **BEVEL PINION SHAFT.** The transmission bevel pinion shaft (35—Fig. CS213) is available only in a

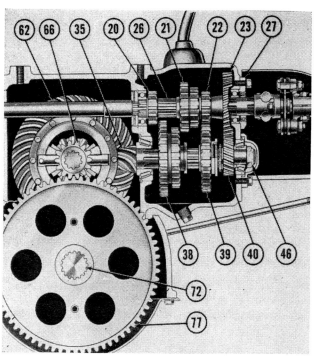

Fig. CS211 – Sectional view of model 20 transmission, differential and final drive.

20. First gear
21. Third gear
22. Second and reverse gear
23. Fourth gear
26. Mainshaft
27. Bearing cap
35. Bevel pinion shaft
38. First and third sliding gear
39. Second and reverse sliding gear
40. Fourth idler gear
46. Adjusting nut
62. Bevel ring (crown) gear
66. Final drive (bull) pinion
72. Rear wheel axle shaft
77. Final drive (bull) gear

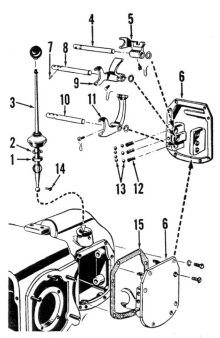

Fig. CS212—Exploded view of model 20 shifter rails and forks.

1. Washer
2. Snap ring
4. Reverse rail
5. Reverse fork
6. Control cover
7. Interlock pin
8. Second and fourth rail
9. Second and fourth fork
10. First and third rail
11. First and third fork
12. Lock spring
13. Lockball
14. Pivot pin
15. Gasket

matched set with the main drive bevel ring gear. If the bevel pinion shaft is damaged, renew both bevel gears. The fore and aft position of the main drive bevel pinion is fixed and non-adjustable.

To remove the bevel pinion shaft, first remove the case side cover and proceed as follows: Remove cover (47) from front of transmission case. Unstake and remove nut (46). Using a lead hammer or soft drift, bump the shaft rearward and withdraw the gears through the side cover opening. The need and procedure for further disassembly is evident after an examination of the unit. After gears are removed, inspect the bushing in the fourth gear (40). The suggested clearance between the pinion shaft and the bushing is 0.0015-0.0035. If the clearance is excessive, renew the bushing and ream the bushing after installation to an inside diameter of 1.0645-1.0655. The pinion shaft diameter at this bushing is 1.062-1.063.

During installation make certain that grooved side of washer (49) is toward gear (40).

When the bevel pinion shaft is installed properly, the taper roller bearings should have a pre-load of 0.002. Obtain the desired pre-load as follows: Install the bevel pinion shaft by reversing the removal procedure and install enough shims (41) to give the shaft some end play. Measure the shaft end play and remove a total thickness of shims equal to 0.002 more than the measured end play. For example, if the measured end play is 0.006, then remove 0.008 thickness of shims. After the adjustment is complete, tighten nut (46) securely and lock the nut by staking a portion of the nut into the shaft groove.

240. **MAIN SHAFT.** The transmission main shaft (26—Fig. CS213) can be removed after removing the reverse idler gear. Remove bearing cap (27) from front of transmission case and working through the rear opening in the case, bump the main shaft forward and out of the gears and case. Remove gears through side cover opening. The need and procedure for further disassembly is evident.

When reinstalling the shaft, vary the number of shims (28) to give the shaft zero end play without causing binding.

Model 30

The standard model 30 transmission is of the four speed type and the transmission gears, differential, main drive bevel pinion and ring gear (crown gear) and the final drive (bull) gears are all contained in the same case. A wall in the case separates the transmission compartment from the differential and final drive compartment. As optional equipment, a set of creeper gears can be installed in the rear section of the intermediate case and the transmission will have eight forward speeds.

Model 30 tractors can be equipped with either a continuous (live) or non-continuous power take-off. The non-continuous power take-off receives its drive from the rear of the transmission main shaft. The drive for the continuous (live) power take-off originates from a spider bolted to the engine clutch cover and is not dependent on any of the transmission shafts or gears.

243. **DETACH (SPLIT) TRANSMISSION CASE FROM INTERMEDIATE CASE.** To split the transmission case from the intermediate case, first support both halves of tractor separately and disconnect the clutch operating rod and tail light wires. On non-Diesel models, remove battery cover, disconnect battery cables and remove battery and battery box. On all models, remove the cover from the intermediate case and on models so equipped, disconnect the belt pulley shifter. Remove the cap screws and nuts retaining the transmission case to the intermediate case and separate the tractor halves.

244. **BASIC PROCEDURES.** Preliminary data for removing the various transmission components are given in the following paragraphs.

245. *Creeper Gears.* The creeper gears (28, 34 & 43—Fig. CS215) can be removed after splitting the transmission case from the intermediate case as outlined in paragraph 243.

246. *Shifter Rails and Forks.* The transmission shifter rails and forks which are attached to transmission case top cover (5—Fig. CS215) can be removed after removing the cover. The procedure for removing the cover is evident.

247. *Reverse Idler Gear.* The reverse idler gear and shaft can be removed after removing the transmission case top cover and splitting the transmission case from the intermediate case as outlined in paragraph 243.

248. *Main Shaft.* The transmission main shaft (D—Fig. CS215) can be removed after splitting the transmission case from the intermediate case and removing the transmission case top cover. Also, remove the rear cover or power take-off unit.

249. *Bevel Pinion Shaft.* To remove the transmission bevel pinion shaft (65—Fig. CS215), first remove the the differential as outlined in paragraph 277; then, split transmission case from the intermediate case as in paragraph 243. Remove the main shaft.

250. **MAJOR OVERHAUL.** A detailed procedure for removing and overhauling the various transmission components is given in the following para-

Fig. CS213—Exploded view of model 20 transmission shafts and gears. Bevel pinion shaft bearings are adjusted with nut (46).

17. Oil cup
18. Snap ring
19. Bearing
20. First gear
21. Third gear
22. Second and reverse gear
23. Fourth gear
24. Bearing
25. Gasket
26. Mainshaft
27. Bearing cap
28. Shims
29. Oil seal
30. Snap ring
31. Reverse idler gear
32. Bushing
33. Reverse idler shaft
34. Lock plate
35. Bevel pinion shaft
36. Bearing cone
37. Bearing cup
38. First and third sliding gear
39. Second and reverse sliding gear
40. Fourth idler gear
41. Shims
42. Bearing cup
43. Snap ring
44. Bearing cone
45. Washer
46. Adjusting nut
47. Bearing cap
48. Gasket
49. Spacer
50. Bushing
51. Snap ring

COCKSHUTT 20-30-40-50   Paragraphs 250-252

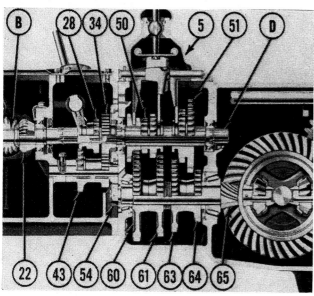

B. Clutch shaft
D. Mainshaft
5. Shifter tower
22. Belt pulley drive pinion
28. Creeper sliding gear
34. Creeper driven gear
43. Creeper idler gear
50. First and second sliding gear
51. Third and fourth sliding gear
54. Adjusting nut
60. First gear
61. Second gear
63. Third gear
64. Fourth gear
65. Bevel pinion shaft

Fig. CS215—Sectional view of model 30 transmission. The bevel pinion and ring (crown) gear are available as a matched unit only.

Fig. CS216—Removing the creeper drive gear retaining snap ring from front end of model 30 transmission mainshaft.

E. Mainshaft
33. Snap ring
34. Creeper driven gear
35. Bearing retainer
54. Adjusting nut
65. Bevel pinion shaft

graphs. Before attempting to overhaul the transmission, perform the following preliminary work. Remove the main drive bevel ring gear (crown gear) and differential assembly as outlined in paragraph 277. Split transmission case from the intermediate case as in paragraph 243.

251. CREEPER GEARS. The creeper driven gear (34—Fig. CS216 or 217) can be removed from front end of the transmission main shaft after removing the retaining snap ring as shown in Fig. CS216. Remove lock wire and set screw retaining the creeper gear shifter fork to the shifter lever shaft and unbolt the shift lever quadrant from right side of intermediate case. Withdraw the shifter lever shaft and remove the shifter fork. Withdraw the sliding gear (28) from the clutch shaft. Remove the lock wire and set screw retaining the creeper idler gear shaft (41) in the intermediate case, bump the shaft forward and withdraw the idler gear (43) from above.

The overhaul procedure for the creeper gears is evident and reassembly is the reverse of the disassembly procedure.

252. SHIFTER RAILS AND FORKS. The shifter rails and forks can be overhauled after removing the transmission case top cover (shifter tower). The procedure for disassembling and overhauling the shifter rails and forks is conventional and readily evident after an examination of the unit and reference to Fig. CS218.

B&C. Clutch shaft
D&E. Transmission mainshaft
21. Snap ring
22. Belt pulley drive pinion
23. Bearing retainer (used prior to Ser. No. 25461)
24. Bearing retainer bolt (used after Ser. No. 25460)
25. Bearing
26. Snap ring
28. Creeper sliding gear
29. Oil seal
30. Snap ring
31. Pilot bearing
32. Sleeve
32A. Bearing
33. Snap ring
33A. Washer
34. Creeper driven gear
35. Bearing retainer
36. Felt seal
37. Washer
38. Seal ring
39. Spacer
40. Thrust washer
41. Idler shaft
42. Bearings
43. Creeper idler gear
44. Washer
45. Bearing spacer
46. Soft plug

Fig. CS217—Exploded view of model 30 clutch shaft, creeper gears and associated parts. Shafts (B) and (D) are used on tractors prior to Ser. No. 3151. Shafts (C) and (E) are used on later models.

CS-63

253. REVERSE IDLER GEAR. To remove the reverse idler gear (49—Fig. CS219) and shaft on tractors after Ser. 25555, first remove the safety wire and the cap screw which retains the reverse idler gear shaft (47) in the transmission case. Withdraw the shaft from front of case and remove the gear from above. On earlier models, the shaft was positioned by a Woodruff key in front end of shaft and the removal procedure is evident. The idler gear bushing (48) can be renewed at this time. Ream the bushing after installation to an inside diameter of 1.127-1.128. The 1.124-1.125 diameter reverse idler shaft should have a clearance of 0.002-0.004 in the bushing.

254. MAIN SHAFT. To remove the transmission main shaft (D or E—Fig. CS219), first remove snap ring as shown in Fig. CS216 and remove gear (34) (or spacer on four speed units) from front end of shaft. Remove the main shaft front bearing retainer (35—Fig. CS219) and using a soft drift against rear end of main shaft, bump the shaft forward until the front bearing emerges from the case. Tilt front of shaft down and working through opening in top of transmission case, remove the snap ring (53) which retains the rear bearing on the mainshaft. Withdraw the shaft forward and remove the gears from above. The need for further disassembly will be readily evident after an examination of the unit. On models prior to serial number 3151, check the pilot bearing (31) in the forward end of the main shaft. On later models, the pilot bearing is located in the rear end of the clutch shaft.

Reinstall the shaft by reversing the removal procedure and make certain that shielded side of both bearings (32A and 52) face toward front of tractor.

255. BEVEL PINION SHAFT. The transmission bevel pinion shaft (65—Fig. CS219) is available only in a matched set with the main drive bevel ring gear. If the bevel pinion shaft is damaged, renew both bevel gears. The fore and aft position of the main drive bevel pinion is fixed and non-adjustable. To remove the bevel pinion shaft and gears, proceed as follows:

Remove the main shaft as in paragraph 254. Remove adjusting nut (54) from front end of bevel pinion shaft, and using a soft drift, bump the shaft rearward and out of transmission case. Withdraw gears from above. The need and procedure for further disassembly is evident after an examination of the unit.

To install the bevel pinion shaft and gears, proceed as follows: Install the gears and spacer spring on a special aligning tool (Cockshutt Part No. TO-9809) and place the assembly in a vise. Compress the assembly as far

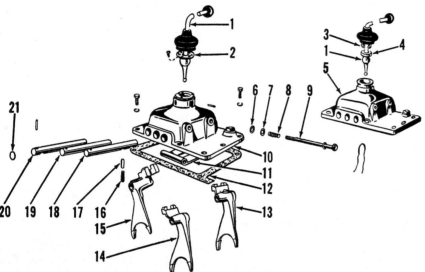

Fig. CS218—Exploded view of model 30 shifter tower, shifter rails and forks.

1. Shifter lever (late models)
2. Lever retainer (late models)
3. Snap ring (early models)
4. Washer (early models)
5. Shifter tower (early models)
6. Lock washer
7. Spring
8. Spring
9. Interlock pin
10. Shifter tower (late models)
11. Interlock plate
12. Gasket
13. First and second fork
14. Third and fourth fork
15. Reverse fork
16. Lock spring
17. Lock pin
18. First and second shaft (rail)
19. Third and fourth shaft (rail)
20. Reverse shaft (rail)
21. Soft plug

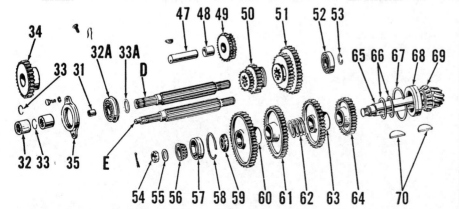

Fig. CS219—Exploded view of model 30 transmission shafts and gears. Gear (34) is only used when tractor is equipped with eight forward speeds.

31. Pilot bearing
32. Sleeve
32A. Bearing
33. Snap ring
33A. Washer
34. Creeper driven gear
35. Bearing retainer
47. Reverse idler shaft
48. Bushing
49. Reverse idler
50. First and second sliding gear
51. Third and fourth sliding gear
52. Bearing
53. Snap ring
54. Adjusting nut
55. Washer
56. Bearing cone
57. Bearing cup
58. Snap ring
59. Washer
60. First gear
61. Second gear
62. Spring
63. Third gear
64. Fourth gear
65. Bevel pinion shaft
66. Washers
67. Snap ring
68. Bearing cup
69. Bearing cone

Fig. CS220—Using heavy wire to hold model 30 transmission drive gears and spring in the compressed position. This procedure is used to facilitate installation of the bevel pinion shaft and gears.

60. First gear
61. Second gear
63. Third gear
64. Fourth gear

COCKSHUTT 20-30-40-50

as it will go without forcing and using a piece of welding rod or heavy wire, wire the gears in the compressed position as shown in Fig. CS220. If the special aligning tool is not available, obtain a piece of suitable bar stock (material C1030 or equivalent) and using the dimensions shown in Fig. CS221, make up an aligning tool.

Note: It is possible to install the bevel pinion shaft and gears without the use of an aligning tool; however, considerably less time will be involved if an aligning tool is used.

After the gears are wired in the compressed position, place the assembly in the transmission case and start the bevel pinion shaft in the gears. Bump the shaft forward and into position, thereby forcing the dummy aligning tool forward and out of the gears. Install the washer and adjusting nut on front end of shaft and pre-load the shaft bearings by tightening the adjusting nut. The adjusting nut should be tightened to give a drag on the taper roller bearings equal to 10-12 In.-Lbs. torque.

Models 40-50

The transmission gears and shafts are all contained in the transmission case as shown in Fig. CS226. The belt pulley unit, when used, is mounted on top of the transmission case and is driven by the high and low range drive gear. The power take-off unit is of the continuous (live) type, driven from a coupling on the engine flywheel, and is not dependent on any of the transmission shafts or gears.

258. **REMOVE AND REINSTALL.** To remove the transmission assembly from tractor, first split the engine frame from the transmission case as outlined in paragraph 217; then, proceed as follows: Support final drive housing and transmission housing separately and remove the transmission shift lever and shifter shaft locking assembly from top of final drive case. Remove the transmission case top cover and the brake and clutch pedals assembly. Unbolt the transmission housing from the final drive housing and separate the units.

Reinstall the transmission by reversing the removal procedure.

259. **BASIC PROCEDURES.** Preliminary data for removing the various transmission components are given in the following paragraphs.

260. *High and Low Range Drive Gear.* The high and low range drive gear (25—Fig. CS226) can be removed after removing the transmission as outlined in paragraph 258 and removing the transmission input shaft front bearing cap and sleeve and the clutch shaft. On models not equipped with the accessory drive shaft, the high and low range drive gear can be removed after splitting engine frame from transmission case.

261. *Countershaft.* The countershaft (43—Fig. CS226) can be removed after removing the transmission and the high and low range drive gear. If the high and low range sliding gear (44) is to be removed, the shifter fork for same must first be removed.

262. *Shifter Shafts and Forks.* The shifter shafts and forks can be removed after removing the transmission and the high and low range drive gear.

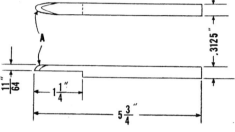
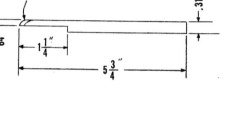

Fig. CS221—Dimensions for making an aligning tool which is used to install the model 30 transmission drive gears and bevel pinion shaft. Material is SAE C1030 or equivalent.

A. Grind off approximately as shown
B. 1⅜ inch counter bore
C. 11/16 inch diameter drill

Fig. CS222—Tightening the model 30 transmission bevel pinion shaft nut.

E. Mainshaft
33. Snap ring
34. Creeper driven gear
35. Bearing retainer

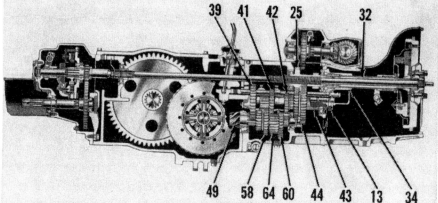

Fig. CS226—Sectional view of models 40 and 50 clutch, transmission and differential. The bevel pinion and bevel ring (crown) gear are available in a matched set only.

13. Shifter shaft guide
25. High and low range drive gear
32. Clutch shaft
34. Cap and sleeve
39. Fourth and sixth gear
41. Second and fifth gear
42. Gear, 1st & 3rd
43. Counter shaft
44. High and low range sliding gear
49. Bevel pinion shaft
58. Second, fourth, fifth and sixth sliding gear
60. First and third sliding gear

263. Reverse Idler Gear and Shaft. The reverse idler gear and shaft can be removed after removing the countershaft as in paragraph 261.

264. Bevel Pinion Shaft. The bevel pinion shaft (49—Fig. CS226) can be removed after removing the countershaft as in paragraph 261.

265. MAJOR OVERHAUL. A detailed procedure for removing and overhauling the various transmission components is given in the following paragraphs. Before attempting to overhaul the transmission, perform the following preliminary work. Remove the transmission as outlined in paragraph 258. Remove the transmission input shaft front bearing cap and sleeve and clutch shaft.

266. HIGH AND LOW RANGE DRIVE GEAR. The high and low range drive gear (25—Fig. CS229) can be removed through the top opening in the case without any further disassembly, and the overhaul procedure is evident. Note: Bearing (26) should be installed with shielded side toward front of tractor and the shield should be removed from bearing (24).

267. COUNTERSHAFT. To remove the countershaft (43—Fig. CS229), first remove the high and low range drive gear. Remove cap screws and bearing washer (36) from rear end of countershaft. Remove the cover clamp (67) from front of case and using a screw driver, pry the countershaft cover (66) from the case bore. Using a soft drift, bump the countershaft forward and remove the fourth and sixth speed drive gear (39), second and fifth speed drive gear (41) and the first and third gear (42). Remove the bearing retaining snap ring from front of shaft, bump the shaft rearward and out of the front bearing. Withdraw the shaft from rear of transmission case. If the high and low range sliding gear (44) is in good condition, no further disassembly is necessary. If, however, the gear must be renewed, it is necessary to remove the high and low range shifter fork.

To install the countershaft, reverse the removal procedure, make certain that beveled side of washer (38) is toward bearing (37) and use Permatex or equivalent around the front cover (66) to seal against possible leaks.

268. SHIFTER SHAFTS AND FORKS. To remove the shifter shafts and forks, first remove the high and low range drive gear (25—Fig. CS229). Remove the shifter shaft guide cap (18—Fig. CS228). Remove the lock wire and set screws retaining the shifter forks to shifter shafts. Bump the shifter shafts rearward and out of case and remove forks from above. The shifter shaft guide can be removed from front of transmission case at this time. Overhaul of the shifter shafts and forks is conventional.

In general, the shifter shafts and forks can be installed by reversing the removal procedure. The following assembly sequence, however, may be helpful.

1. Lay the second and third fork (9) in position.
2. If the high and low range sliding gear (44—Fig. CS229) has been removed, place the gear in position at this time.
3. Lay the high and low fork (19—Fig. CS228) into position.
4. Install the shifter shaft guide (13).
5. Install the high and low shifter shaft (20).
6. Install both shift lock balls (18A).
7. Install interlock pin (16).
8. Install second and third shifter shaft (8).
9. Lay the first and reverse fork (11) in position.
10. Install and safety wire the high and low shifter fork set screw.
11. Place gears in high speed position and install first and reverse shifter shaft (10).
12. Install lock pins (15), springs (14) and plugs. CAUTION: The plugs should be tightened to give a positive holding action in any gear position.
13. Install the guide cover (18).
14. Install and safety wire the remaining shifter fork set screws.

269. REVERSE IDLER GEAR AND SHAFT. The reverse idler gear (64—Fig. CS229) and shaft can be removed after removing the countershaft as outlined in paragraph 267. Remove both lock screws (68 and 69) from right side of transmission case. Note: The second screw (68) is accessible after removing the first screw. Withdraw the reverse idler shaft (55) from rear and remove gear from above. Renew the idler shaft and/or roller bearing if either is excessively worn.

270. BEVEL PINION SHAFT. The transmission bevel pinion shaft (49—Fig. CS229) is available only in a matched set with the main drive bevel ring gear. If the bevel pinion shaft is damaged, renew both bevel gears.

The fore and aft position of the main drive bevel pinion is fixed and non-adjustable. To remove the bevel pinion shaft and gears, proceed as follows:

Remove the countershaft as in paragraph 267. Using a screw driver, pry the pinion shaft front cover (66) from the case bore. Remove the adjusting nut (63) and washer and using a soft drift, bump the pinion shaft rearward and withdraw gears from above.

Install the main drive bevel pinion shaft by reversing the removal procedure and pre-load the taper roller bearings by tightening the adjusting nut (63). The adjusting nut should be tightened to give a drag on the taper roller bearings equal to 10-12 In.-Lbs. torque. Use Permatex or equivalent around the pinion shaft front cover to seal against possible leaks.

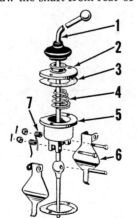

Fig. CS227—Exploded view of models 40 and 50 shifter lever assembly.

1. Shifter lever
2. Cap
3. Gasket
4. Spring
5. Shifter tower
6. Overshift lock
7. Interlock torsion spring

8. Second and third shaft
9. Second and third fork
10. First and reverse shaft
11. First and reverse fork
12. Gasket
13. Shifter shaft guide
14. Lock spring
15. Lock pin
16. Interlock pin
17. Gasket
18. Cover
18A. Lock balls
19. High and low range fork
20. High and low range shaft
21. Shift block

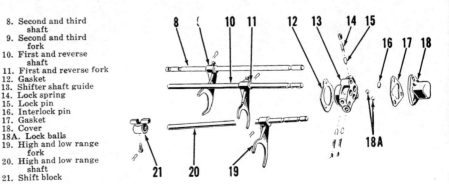

Fig. CS228—Models 40 and 50 shifter shafts and forks.

COCKSHUTT 20-30-40-50

Paragraph 273

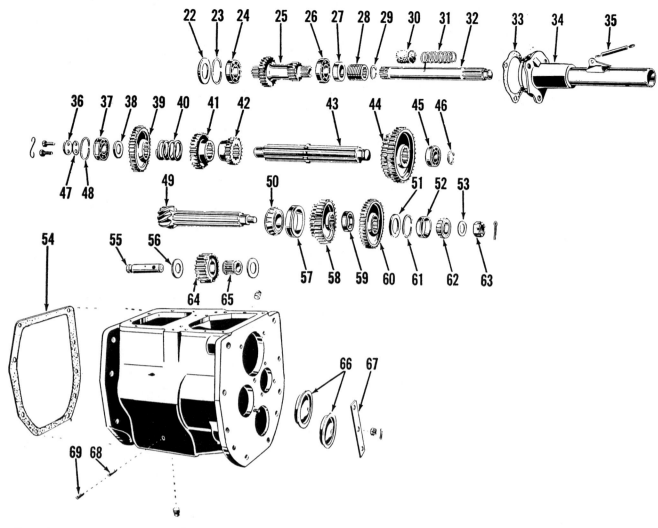

Fig. CS229—Exploded view of models 40 and 50 transmission. The bevel pinion shaft bearings are adjusted with nut (63).

22. Bearing cover
23. Snap ring
24. Bearing
25. High and low range drive gear
26. Bearing
27. Spacer
28. Collar
29. Snap ring
30. Felt seal
31. Spring
32. Clutch shaft
33. Gasket
34. Bearing cap and sleeve
35. Spring
36. Washer
37. Bearing
39. Fourth and sixth gear
40. Spring
41. Second and fifth gear
42. First and third gear
43. Counter shaft
44. High and low range sliding gear
45. Bearing
46. Snap ring
47. Shim
48. Snap ring
49. Bevel pinion shaft
50. Bearing cone
51. Cover
52. Bearing cup
53. Washer
54. Gasket
55. Reverse idler shaft
56. Thrust washer
57. Bearing cup
58. Second, fourth, fifth and sixth sliding gear
59. Spacer
60. First and third sliding gear
62. Bearing cone
63. Adjusting nut
64. Reverse idler gear
65. Bearing
66. Covers
67. Cover clamp
68. Inner lock screw
69. Outer lock screw

MAIN DRIVE BEVEL GEARS AND DIFFERENTIAL

ADJUST BEVEL GEARS AND DIFFERENTIAL CARRIER BEARINGS
Models 20-30-40-50

The fore and aft position of the main drive bevel pinion is fixed and non-adjustable. The backlash between the main drive bevel ring gear (crown gear) and the main drive bevel pinion is adjusted by shims (A—Fig. CS231, 232, or 233). These same shims (A) also control the differential carrier bearing adjustment. The subsequent adjustment procedure will be outlined on the assumption that a new set of bevel gears have been installed and that the interfering parts are out of the way so that the adjusting shims are accessible.

273. The recommended procedure in adjusting a set of bevel gears is to place all of the shims (A—Figs. CS231, 232 or 233) on one side so that the bevel gears have maximum backlash. Then, add or remove shims until the differential carrier bearings have a pre-load equal to 10-12 In.-Lbs. torque. At this time, the bevel gears

will still have maximum backlash. Now, transfer as many shims as necessary to the other side to obtain the desired bevel gear backlash. The backlash should be 0.008-0.012 on all models. Shims are available in thicknesses of 0.004 and 0.010 for model 20, 0.003, 0.005 and 0.010 for models 30, 40 and 50.

RENEW BEVEL GEARS
Models 20-30-40-50

274. The main drive bevel pinion gear which is integral with a transmission shaft and the main drive bevel ring gear (crown gear) are available in matched sets ONLY. Therefore, if either the pinion or the ring gear is damaged, it will be necessary to renew both bevel gears as follows:

To renew the bevel pinion and integral shaft, follow the procedure outlined in paragraph 239 for model 20, paragraph 255 for model 30 and paragraph 270 for models 40 and 50. The main drive bevel ring (crown) gear on models 20 and 30 can be renewed at this time. The model 20 ring gear is retained to the differential case by rivets; whereas, the model 30 ring gear is retained by safety-wired cap screws. Models 40 and 50 bevel ring gear, which is riveted to the differential case, can be renewed after removing the differential as outlined in paragraph 282.

After a new set of bevel gears has been installed, adjust the differential carrier bearings and the main drive bevel gear backlash as outlined in paragraph 273.

DIFFERENTIAL
Model 20

The differential is of the two pinion type as shown in Fig. CS233. The main drive bevel ring gear (crown gear) is retained to the one piece case by twelve rivets. The differential and bevel ring gear assembly is mounted back of a dividing wall in the transmission case.

275. **REMOVE AND REINSTALL.** To remove the differential assembly, first remove both of the final drive housing assemblies as outlined in paragraph 285. Remove the transmission case rear cover or combination belt pulley and power take-off unit. Remove both of the differential bearing carriers (54—Fig. CS233) and withdraw the differential and bevel ring gear assembly from rear of case. Note: Be careful not to mix or lose shims (A) which are installed between the bearing carriers and the transmission case.

When reinstalling the differential, adjust the differential carrier bearings and check (and adjust if necessary) the main drive bevel gear backlash as outlined in paragraph 273.

276. **DISASSEMBLE AND OVERHAUL.** After the differential is removed as outlined in paragraph 275, proceed to overhaul the differential as follows: Using a thin drift punch, drive out the pinion lock pin (64—Fig. CS233). Bump the pinion pin (57) out of the differential case and remove pinions (59), thrust washers (58), side gears (60) and thrust washers (61). The taper roller carrier bearing cones can be pulled from the differential case and the bearing cups can be driven from the bearing carriers. Inspect the oil seal (53) in each bearing carrier. Always renew a questionable seal. Renew thrust washers (58 or 61) if they show wear.

Reassemble the differential by reversing the disassembly procedure and lock the pinion pin lock pin (64) in place by burring sides of hole with a punch.

Model 30

The differential is of the two pinion type. The main drive bevel ring (crown gear) is retained to the one piece differential case by twelve safety wired cap screws. The differential and bevel ring gear assembly is mounted back of a dividing wall in the transmission case.

277. **REMOVE AND REINSTALL.** To remove the differential and main drive bevel ring gear assembly, first remove both of the final drive (bull) pinions and differential side gear units as outlined in paragraph 293; then, remove the differential and bevel ring gear assembly through the rear open-

Fig. CS231—Sectional view of model 30 differential and final drive.

A. Shims	5. Bearing cup	14. Bevel ring gear	18. Bull gear
1. Bull pinion and differential side gear	6. Oil seal	15. Differential case and cross shaft	19. Axle shaft
	7. Bearing cap		20. Bearing
4. Bearing cone	10. Pinion	16. Nut	21. Axle housing (trumpet)
	12. Pinion pin lock pin	17. Locking washer	

COCKSHUTT 20-30-40-50

ing in the transmission case. Frequently, difficulty is encountered when attempting to withdraw the differential assembly. In which case, the following suggested procedure may be helpful.

Move differential toward right until the cross shaft extends through the right hand bore as shown in Fig. CS234. With left hand inserted through the transmission case rear opening, grasp the cross shaft to left of bevel gear. With right hand outside the case, grasp the protruding right end of the cross shaft. Raise left end of cross shaft out of bore on left side of case and rotate left end of cross shaft upward until right end of cross shaft can be moved forward and out

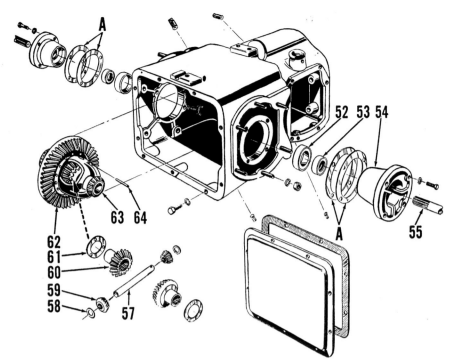

3. Differential case right half
4. Side gear
6. Bevel ring gear
7. Pinion
9. Differential case left half
12. Bearing cage
16. Bull pinion shaft
17. Nut
18. Bearing cup
19. Bearing cone
20. Bull gear
21. Axle shaft
21A. Oil seal
22. Axle housing (trumpet)
49. Bevel pinion shaft

Fig. CS233—Exploded view of model 20 differential and associated parts. The differential carrier bearings and bevel gear backlash are adjusted with shims (A).

52. Bearing cup
53. Oil seal
54. Bearing carrier
55. Bull pinion shaft
57. Pinion pin
58. Thrust washer
59. Pinion
60. Side gear
61. Thrust washer
62. Bevel ring gear
63. Bearing cone
64. Lock pin

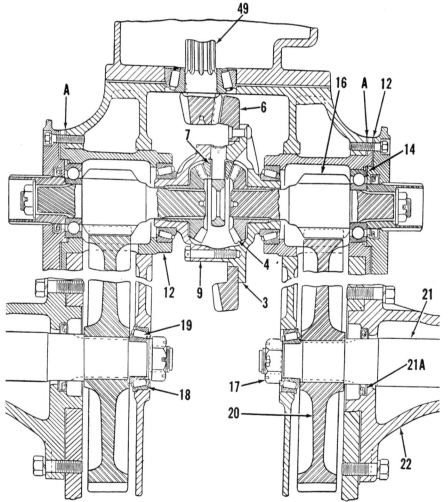

Fig. CS232—Sectional view of models 40 and 50 differential and final drive. The differential carrier bearings and bevel gear backlash are adjusted with shims (A).

of the right hand case bore. Now allow the unit to come to rest as shown in Fig. CS235. Grasp left end of cross shaft with left hand and grasp the bevel ring gear with the right hand. Rotate left end of cross shaft forward, and at the same time, raise the differential as high as possible. With differential in the raised position, move the right end of cross shaft rearward just beyond the web (or rib) in case as

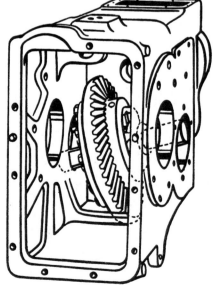

Fig. CS234—First step in withdrawing model 30 differential.

shown in Fig. CS236. Raise the right end of the cross shaft so that it enters the right rear axle bore in the case and bring left end of cross shaft toward rear as shown in Fig. CS237. Bring left end of cross shaft out through rear opening in case as in Fig. CS238 and remove the differential assembly.

Note: The differential cross shaft is a tight fit in the differential case; however there have been a few isolated cases where the cross shaft has moved in the case and it is impossible to remove the differential by following the above procedure. When these isolated cases are found, it will be necessary to drive or press the cross shaft back into position—centered in differential case.

When installing the differential, reverse the removal procedure and adjust the differential carrier bearings and check (and adjust if necessary) the main drive bevel gear backlash as outlined in paragraph 273.

280. **DISASSEMBLE AND OVERHAUL.** After the differential is removed as outlined in paragraph 277, proceed as follows: Using a thin drift punch, drive out both of the pinion pin lock pins (12—Fig. CS239) and withdraw pinions (10) and thrust washers (9). Check all parts and renew any which are excessively worn. The pinions on some early production tractors were not fitted with bushings. Only the late production, bushing type pinions are available for replacement. When installing new bushings (11), ream them after installation to an inside diameter of 1.002-1.004. The 0.997-0.998 diameter pinion shafts should have a clearance of 0.004-0.007 in the bushings.

Reassemble the differential by reversing the disassembly procedure and lock the pinion pin lock pins (12) in place by burring sides of hole with a punch.

Models 40-50

The differential is of the four pinion type as shown in Fig. CS242. The main drive bevel ring (crown) gear is retained to the right half of the differential case by twelve rivets. The differential assembly is mounted in the center section of the final drive case.

282. **REMOVE AND REINSTALL.** To remove the differential and main drive bevel ring gear assembly, remove both

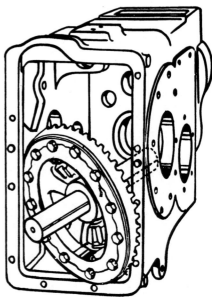

Fig. CS238—Fifth step in withdrawing model 30 differential.

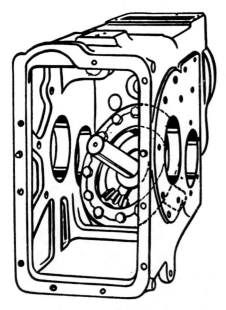

Fig. CS235—Second step in withdrawing model 30 differential.

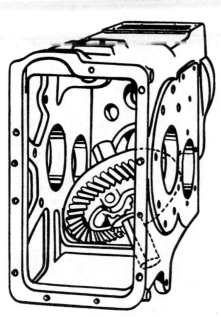

Fig. CS236—Third step in withdrawing model 30 differential.

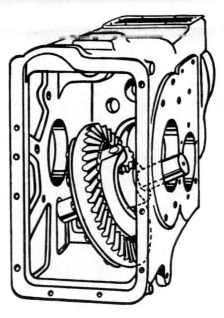

Fig. CS237—Fourth step in withdrawing model 30 differential.

Fig. CS238A—Model 30 final drive bull gear and pinion. The pinion is integral with the differential side gear.

COCKSHUTT 20-30-40-50

Paragraphs 282-283

Fig. CS238B — Model 30 main drive bevel pinion and ring (crown) gear. The gears are available only in a matched set.

Fig. CS243—Rear view of models 40 and 50 final drive case with PTO housing removed.

R. Bearing retainer T. Spring
P. Cap screw S. Oiler angle iron

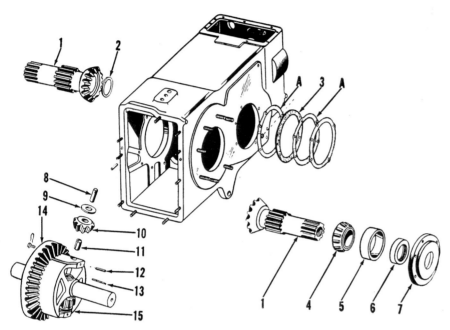

Fig. CS239—Exploded view of model 30 differential and associated parts. The differential carrier bearings and bevel gear backlash are adjusted with shims (A).

1. Bull pinion and differential side gear
2. Thrust washer
3. Gasket
4. Bearing cone
5. Bearing cup
6. Oil seal
7. Bearing cap
8. Pinion pin
9. Thrust washer
10. Pinion
11. Bushing
12. Locking pin
13. Pin
14. Bevel ring gear
15. Differential case and shaft

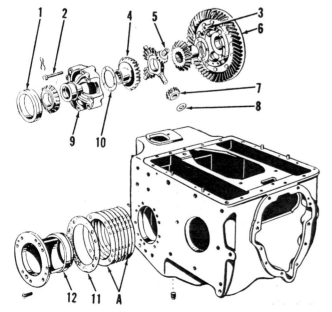

Fig. CS242 — Exploded view of models 40 and 50 differential assembly.

1. Bearing cup
2. Bolt
3. Differential case right half
4. Side gear
5. Spider
6. Bevel ring gear
7. Pinion
8. Thrust washer
9. Differential case left half
10. Thrust washer
11. Gasket
12. Bearing cage

of the final drive (bull) gears as outlined in paragraph 296 and both of the final drive (bull) pinions as outlined in paragraph 295. On very late models, remove the hydraulic pump drive gear shifting mechanism. Remove the final drive case rear cover or power take-off unit. Remove the cap screws (P—Fig. CS243) retaining the accessory drive shaft in the final drive case and withdraw the drive shaft. On top of the final drive case, locate and loosen the vertical stud which positions the power take-off oiler angle iron (S) in the case and remove the angle iron. Remove both of the differential bearing cages (12—Fig. CS242) and be careful not to mix or lose shims (A) which are located under the cages. Lift differential and bevel ring gear assembly from the final drive case.

When installing the differential, reverse the removal procedure and adjust the differential carrier bearings and check (and adjust if necessary) the main drive bevel gear backlash as outlined in paragraph 273. The oiler angle iron (S—Fig. CS243) should be installed with cut-out toward rear of tractor so as to allow clearance for spring (T).

283. **DISASSEMBLE AND OVERHAUL.** After the differential is removed as outlined in paragraph 282, proceed as follows: Remove the eight bolts (2—Fig. CS242) joining the two halves of the differential case and separate the two halves. Disassemble the remaining parts and inspect them for damage or wear. Renew any questionable parts.

Reassemble the differential by reversing the disassembly procedure.

FINAL DRIVE
Model 20

As treated in this section, the final drive will include the final drive housings, the drive (bull) pinions and shafts and the drive (bull) gears and wheel axle shafts.

COCKSHUTT 20-30-40-50

285. REMOVE AND REINSTALL FINAL DRIVE HOUSING ASSEMBLY. To remove either final drive housing assembly, first disconnect light wires and brake rod. Support rear of tractor and remove wheel and tire unit. Remove fender and unbolt the final drive housing from the transmission case. Carefully withdraw the final drive housing straight out from transmission case.

Install the final drive housing assembly by reversing the removal procedure.

286. DRIVE (BULL) PINION AND/OR SHAFT. To remove either of the final drive (bull) pinions and/or shafts, first remove the respective final drive housing from tractor. Remove pinion bearing cap (71—Fig. CS245) and extract snap ring (69). Using a soft drift against outer end of pinion shaft, bump shaft out of bearing (67), pinion (66) and housing. Withdraw the bearing and drive (bull) pinion. The brake drum (65) can be removed from the pinion shaft after removing the drum retaining set screw.

When reassembling the unit, it may be necessary to back-off on the brake adjustment. If this is done, readjust the brakes as outlined in paragraph 300.

287. DRIVE (BULL) GEAR AND/OR WHEEL AXLE SHAFT. The drive (bull) gear and/or wheel axle shaft can be removed without removing the final drive housing from the tractor. Support rear of tractor and remove the wheel and tire unit. Remove inner cover cap (83—Fig. CS245) and gear cover (79). Remove adjusting nut (84) and washer (82). Using a very small cold chisel inserted between inner edge of bull gear and inner surface of the final drive housing, split snap ring (80). Attach a suitable puller as shown in Fig. CS246 and push the wheel axle shaft out of the bull gear. The need and procedure for further disassembly is evident after an examination of the unit.

When reassembling the unit, adjust the taper roller axle bearings by tightening nut (84—Fig. CS245) to provide zero end play without causing excessive binding.

Model 30

As treated in this section, the final drive will include the final drive (bull) gears, the wheel axle shafts and housings and the integral final drive (bull) pinions and differential side gears.

290. DRIVE (BULL) GEARS AND/OR WHEEL AXLE SHAFT. To remove either bull gear, first support rear of tractor and remove the respective wheel and tire unit. Drain the transmission case and remove the

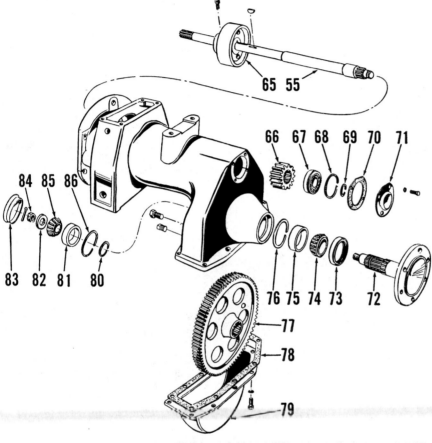

Fig. CS245—Exploded view of model 20 final housing and associated parts. The wheel axle shaft bearings are adjusted with nut (84).

55. Bull pinion shaft	71. Cover	76. Snap ring	81. Bearing cup
65. Brake drum	72. Wheel axle shaft	77. Bull gear	82. Washer
66. Bull pinion	73. Oil seal	78. Gasket	83. Cover cap
67. Bearing	74. Bearing cone	79. Housing cover	84. Adjusting nut
68. Snap ring	75. Bearing cup	80. Snap ring	85. Bearing cone
69. Snap ring			86. Snap ring

Fig. CS246—Puller arrangement for pressing the model 20 wheel axle shaft out of the final drive bull gear.

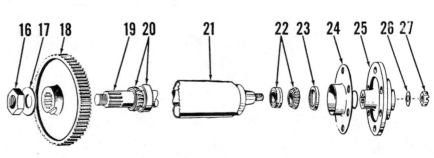

Fig. CS249—Exploded view of model 30 bull gear, wheel axle shaft and associated parts.

16. Nut	19. Axle shaft	22. Outer bearing	25. Hub
17. Locking washer	20. Inner bearing	23. Oil seal	26. Washer
18. Bull gear	21. Axle housing	24. Slinger	27. Nut

seat assembly. Remove the transmission case rear cover or power take-off unit. Remove the platform and fender. Remove the drawbar brace. On models equipped with shoe type brakes, remove the shoes. Working through the rear opening in the transmission case, remove the nut (16—Fig. CS249) retaining the drive (bull) gear to inner end of the wheel axle shaft (19). Unbolt the axle housing (trumpet) from the transmission case and using a heavy bar against the fender mounting pad, bump the axle and housing assembly away from the transmission case. The bull gear can be removed at this time.

Note: On some late model tractors, the drive (bull) gears are tapped to receive puller screws. In which case, the bull gear can be pulled from the inner end of the axle shaft. The puller can be made, using the following specifications.

Obtain a steel bar approximately ½ by 1½ by 7 inches and drill two 9/16 inch holes through the bar. The center-to-center distance between the holes should be 5⅝ inches. Use two ½-13 by 3 inch cap screws through the bar and screw them in to the tapped holes in the bull gear.

The procedure for removing the wheel axle shaft from the axle housing is evident after the bull gear is removed.

When reassembling the unit, tighten the drive (bull) gear nut, as shown in Fig. CS250, enough to give the wheel axle carrier bearings a pre-load of 105 In.-Lbs. torque. Note: It will be possible to rotate the axle shaft only through the backlash between the bull gear and pinion. After the adjustment is complete, lock the nut (16—Fig. CS249) in position by upsetting washer (17) against nut and cut-away in gear.

293. **DRIVE (BULL) PINION.** To remove either of the drive (bull) pinions and integral differential side gears, first remove the respective drive (bull) gear as outlined in paragraph 290. Remove the brake drum. Remove the differential bearing cap as shown in Fig. CS251. Using an impact type puller screwed into the outer end of the bull pinion, remove the pinion (1—Fig. CS252), bearing cup (5) and cone (4). The differential side gear thrust washer should be renewed if it shows wear.

Reinstall the bull pinion by reversing the removal procedure and adjust the brakes as outlined in paragraph 303 or 307.

Models 40-50

As treated in this section, the final drive will include the final drive (bull) gears, the wheel axle shafts and housings and the final drive (bull) pinions.

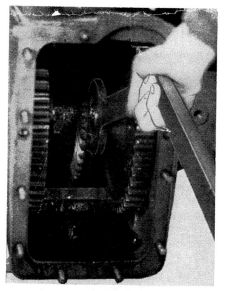

Fig. CS250—Tightening the bull gear retaining nut on model 30. The nut should be tightened enough to give the axle shaft bearings a slight pre-load. See text.

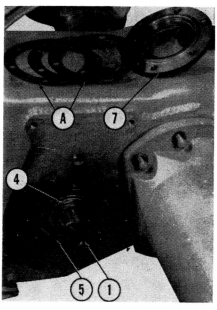

Fig. CS252—Model 30 differential carrier bearings (4 & 5) are adjusted with shims (A) which are located under bearing cap (7). These same shims (A) also control the main drive bevel gear backlash.

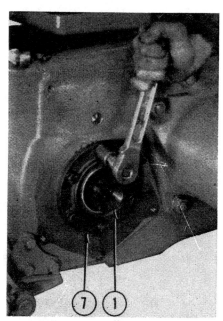

Fig. CS251—Removing model 30 differential carrier bearing cap (7). The bull pinion and integral differential side gear is shown at (1).

Fig. CS254—Cut-away view of models 40 and 50 final drive. For exploded views and identification of parts, refer to Figs. CS242 and 255.

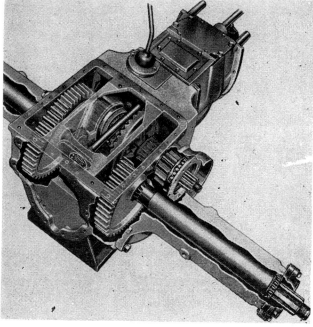

295. DRIVE (BULL) PINIONS. To remove either of the final drive bull pinions (16—Fig. CS255), first disconnect brake rod and remove the brake shoes or discs. On models with shoe type brakes, remove the brake anchor. On models with disc type brakes, remove the brake housing. Withdraw the drive (bull) pinion. The need for further disassembly is evident.

Install the pinions by reversing the removal procedure.

296. DRIVE (BULL) GEARS AND/OR WHEEL AXLE SHAFTS. To remove either bull gear (20—Fig. CS255) and/or wheel axle shaft (21), proceed as follows: Remove seat, final drive case cover and the respective wheel and tire unit. Remove cotter pin and nut retaining the drive (bull) gear to the inner end of the wheel axle shaft. Loosen the cap screws retaining the bearing cap (27) to the outer end of the wheel axle housing (trumpet). Using a heavy bar against the wheel drive flange, bump the wheel axle shaft outward and again loosen the cap screws. Bump the drive flange again and remove the cap screws. Continue bumping the axle shaft outward until the axle is free from the drive (bull) gear. Lift bull gear from the final drive case. The need and procedure for further disassembly is evident.

When reassembling the unit, vary the number of shims (26), which are available in thicknesses of 0.003, 0.005 and 0.010, to give the taper roller axle bearings a pre-load of 0.000-0.002. One method of obtaining this pre-load is to install more than enough shims and measure the axle shaft end play. Then remove a total thickness of shims equal to 0.001 more than the measured end play. For example, if the measured end play is 0.005, remove 0.006 thickness of shims.

Fig. CS255—Exploded view of models 40 and 50 final drive.

A. Shims
12. Bearing cage
13. Snap ring
14. Bearing
15. Gasket
16. Bull pinion
17. Nut
18. Bearing cup
19. Bearing cone
20. Bull gear
21. Axle shaft
21A. Oil seal
22. Axle housing
23. Shield
24. Bearing
25. Gasket
26. Shims
27. Bearing cap
28. Oil seal
29. Dirt shield

contact the adjusting bolts (93) when the top covers are installed and the cover cap screws are securely tightened.

Adjust the pedals by means of the connecting eye bolts (95—Fig. CS261) so that both pedals have an equal amount of free travel.

301. R&R BRAKE SHOES. To remove either set of brake shoes, disconnect the brake pull rod, remove the top cover plate (88—Fig. CS261), remove adjusting bolt (93) and disconnect springs (90). Remove the bottom cover plate (98) and eye bolt (95). Withdraw shoes from above.

When reinstalling the shoes, do not hook springs (90) or tighten adjusting nut (91) until after the bottom cover plate (98) is assembled to the housing. Adjust the brakes as outlined in paragraph 300.

Note: Greasy brake linings are an indication that the differential bearing carrier oil seal has failed. To renew the seal, first remove the final drive housing as outlined in paragraph 285. Remove the differential bearing carrier and renew the seal. Faulty differential carrier bearings can cause an oil seal to fail. It is an approved practice to check for worn bearings at this time.

302. R&R BRAKE DRUM. The procedure for removing the brake drum is evident after removing the drive (bull) pinion shaft as outlined in paragraph 286.

BRAKES

Model 20

As shown in Fig. CS260, the brakes are of the two shoe type that clamp on a brake drum which is keyed to each final drive (bull) pinion shaft.

300. ADJUSTMENT. To adjust the brakes, support rear of tractor and remove upper cover (88—Fig. CS260). Tighten nut (91) on each brake until rear wheels cannot be turned by hand; then, back the adjusting nut off 2½-3 turns or until drag is just removed. The brake shoes are centered with respect to the brake drum by the adjustable brake guides (89) which are mounted on the bottom side of each top cover (88). Test freedom of rotation of each rear wheel after top cover is installed to make certain that guides (89) are not forcing the top ends of the shoes against drums; thus, throwing the shoes eccentric with respect to the drums and causing a rotational drag. The guides should just

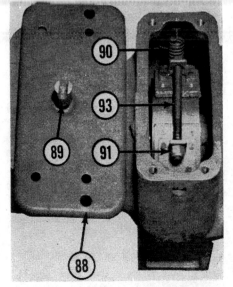

Fig. CS260—Model 20 brakes can be adjusted after removing the housing top cover.

88. Top cover
89. Brake guide
91. Adjusting nut
93. Adjusting screw

COCKSHUTT 20-30-40-50

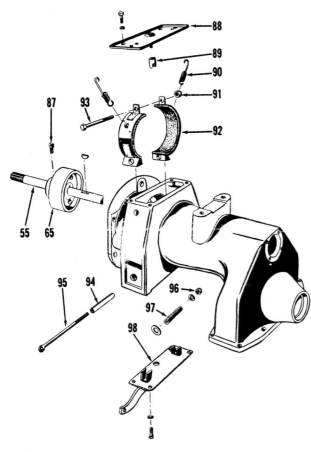

Fig. CS261—Model 20 brake parts exploded from the final drive housing.

55. Bull pinion shaft
65. Brake drum
87. Set screw
88. Top cover
89. Brake guide
90. Spring
91. Adjusting nut
92. Shoes
93. Adjusting screw
94. Brake rod tube
95. Brake rod
96. Nut
97. Spring
98. Bottom cover

Note: Greasy brake linings are an indication that the oil seal in the differential bearing cap has failed. To renew the seal, first remove the brake drum as in paragraph 305 and unbolt the bearing cap from the transmission case. The seal can be renewed at this time. Faulty differential carrier bearings can cause an oil seal to fail. It is an approved practice to check for worn bearings at this time. Also, check for a cracked brake drum hub. Always renew the rubber seal ring.

305. **R&R BRAKE DRUMS.** To renew either brake drum, first remove the band and brake cover assembly as shown in Fig. CS264. Remove the drum retaining cap screw and while prying on the drum from behind, strike the drum with a raw hide hammer and remove the drum. Refer to Fig. CS265.

When reinstalling, always renew the rubber seal ring and tighten the cap screw securely as shown in Fig. CS266.

Models 30-40 (Shoe Type)

Model 30 tractors after Serial 10,000, and all model 40 tractors are factory equipped with clamping type shoe brakes as shown in Figs. CS267 or 269.

307. **ADJUSTMENT.** To adjust the brakes, vary the length of each brake pull rod (35—Fig. CS267 or 269) until pedals have a free travel of 1-1½ inches. To synchronize both brakes, jack up rear end of tractor, start engine and shift the transmission into

Model 30 (Band Type)

Model 30 tractors prior to Serial No. 10,001 were factory equipped with band type brakes as shown in Fig. CS263.

303. **ADJUSTMENT.** To adjust the brakes, support rear of tractor and remove the platforms (floor boards). Remove plate (2—Fig. CS263) from top of each brake cover and tighten adjusting nuts (X) until rear wheels cannot be turned by hand; then, back the nuts off until drag is just removed.

Adjust the brake linkage so that each brake pedal has an equal amount of free travel.

304. **R&R BRAKE BANDS.** To remove the brake bands, remove the platforms (floor boards) and disconnect the operating linkage. Remove the brake covers and band assemblies as in Fig. CS264, and remove the bands.

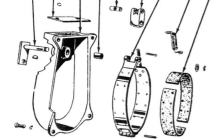

Fig. CS263—Exploded view of early model 30 band type brakes. The brakes are adjusted with nut (X).

1. Lever assembly
2. Housing cover
3. Toggle pin
4. Toggle
5.
6. Spring
7. Linings
8. Bushing
9. Housing

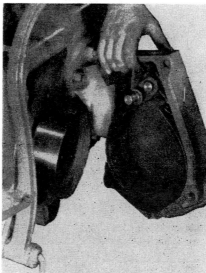

Fig. CS264—Removing model 30 brake housing and band assembly.

Fig. CS265—Removing model 30 brake drum. Model 40 is similar.

fourth gear. Depress both pedals at the same time. If one wheel slows down before the other, loosen the adjustment on the tight brake.

308. **R&R SHOES.** To remove brake shoes, proceed as follows: On model 30 Diesels, first remove battery box and platform. On all models, disconnect brake rods, remove snap rings retaining brake shoes assemblies to anchor pins and remove shoes. When reinstalling, adjust the brakes as outlined in paragraph 307.

Note: Greasy brake linings are an indication that oil seal (37—Fig. CS267) or (24—Fig. CS269) has failed. To renew the seal, first remove the brake drum as shown in Fig. CS265 and unbolt the bearing cap or brake anchor plate from transmission or final drive case. The seal can be renewed at this time.

On model 30, faulty differential carrier bearings can cause an oil seal to fail. On model 40, faulty bull pinion shaft bearings can cause the oil seal to fail. It is an approved practice to check for worn bearings at this time.

On all models, check for a cracked brake drum hub. Always renew the rubber seal ring.

309. **R&R BRAKE DRUM.** After removing the brake shoes, remove the drums as shown in Fig. CS265.

Model 50

Model 50 tractors are equipped with double disc, self-energizing brakes. The brakes are mounted on the outer ends of the bull pinion shafts.

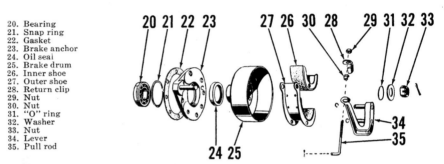

20. Bearing
21. Snap ring
22. Gasket
23. Brake anchor
24. Oil seal
25. Brake drum
26. Inner shoe
27. Outer shoe
28. Return clip
29. Nut
30. Nut
31. "O" ring
32. Washer
33. Nut
34. Lever
35. Pull rod

Fig. CS269—Exploded view of model 40 shoe type brakes.

Fig. CS266—Using a long wrench to tighten the model 30 brake drum cap screw. Model 40 is similar.

Fig. CS271—Model 50 double disc brakes are adjusted with nut (14). Item (15) is a jam nut.

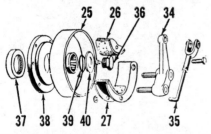

Fig. CS267—Exploded view of model 30 shoe type brakes.

25. Brake drum	36. Cap screw
26. Inner shoe	37. Oil seal
27. Outer shoe	38. Bearing cap
34. Lever	39. "O" ring
35. Pull rod	40. Washer

310. **ADJUSTMENT.** To adjust the brakes, loosen the lock nut and tighten adjusting nut (14—Fig. CS271) to reduce the pedal free travel to not less than one inch. To synchronize both brakes, jack up rear end of tractor, start engine and shift the transmission into fourth gear. Depress both pedals at the same time. If one wheel slows down before the other, loosen the adjustment on the tight brake.

311. **R&R DISCS.** To remove the lined discs, disconnect the brake rods and remove the brake housing covers. Withdraw lined discs and actuating discs.

After brake discs are installed, adjust the pedal travel as in paragraph 310.

Note: Greasy disc linings are an indication that oil seal (2—Fig. CS272) is leaking. The seal can be renewed after removing the brake housing as outlined in paragraph 312. Faulty bull pinion shaft bearings can cause the oil seal to fail. Check the bearings at this time.

312. **R&R BRAKE HOUSINGS.** To remove either brake housing, first remove the lined discs as outlined in paragraph 311. Remove nut (N—Fig. CS272) and extract drive sleeve (5). Unbolt and remove the housing.

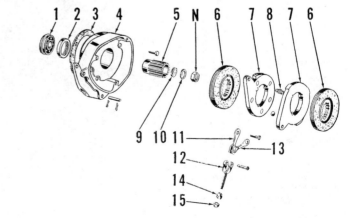

1. Bearing
2. Oil seal
3. Gasket
4. Housing
5. Drive sleeve
6. Lined disc
7. Actuating disc
8. Spring
9. Cork washer
10. Washer
11. Actuating link
12. Yoke
13. Actuating link
14. Adjusting nut
15. Jam nut

Fig. CS272—Exploded view of model 50 double disc brakes. The brakes are adjusted with nut (14).

COMBINED BP & PTO UNIT

Model 20

The combination belt pulley and power take-off unit is mounted on the rear face of the transmission case and is driven by the splined rear section of the transmission main shaft. For removal and overhaul of the transmission main shaft, refer to paragraph 240.

Power to the combination unit is transmitted through drive shaft (1—Fig. CS280), through sliding gear (9) and to the spur teeth of gear (14) which is keyed to the power take-off shaft (36). Power to the belt pulley shaft is supplied by the bevel portion of gear (14) which is in constant mesh with bevel gear (21). Bevel gear (21) is splined to the belt pulley shaft (41).

320. OVERHAUL. The procedure for removing the belt pulley and power take-off unit is self-evident. To disassemble the removed unit, proceed as follows: Withdraw drive shaft (1) and sliding gear (9). Remove the power take-off shaft rear bearing cap (35) and save shims and gaskets (34 and 37) for reinstallation. Attach a suitable puller tool in a manner similar to that shown in Fig. CS281 and press the power take-off shaft rearward and out of gear (14—Fig. CS 280). The shaft front bearing cup (12)

Fig. CS281—Puller hook-up for pressing the power take-off output shaft out of the spur and spiral bevel gear.

and shims (11) will remain in the housing. If this bearing cup is removed for any reason, be certain to mark and keep together the shim or shims (11) which are located behind the bearing cup. These shims are used to control the fore and aft position of the combination spur and bevel gear (14).

To remove the belt pulley shaft, proceed as follows: Remove Welch plug (19), extract the cotter pin and remove nut (20). Bump the belt pulley shaft out of bevel gear (21) and housing. Withdraw the bevel gear and save shims (22) for reinstallation. The need and procedure for further disassembly is evident after an examination of the unit.

In general, the belt pulley and power take-off shafts are installed by reversing the removal procedure. The following, however, should be observed. When installing the belt pulley shaft, use the same thickness and number of shims (22) as were originally removed and tighten nut (20) to remove all shaft end play without causing the taper roller bearings to bind. If the power take-off shaft front bearing cup (12) was removed, use the same number of shims (11) behind the cup as were originally installed. Install the power take-off shaft and vary the number of shims (34) which are available in thicknesses of 0.005 and 0.010 to remove all shaft end play without causing the taper roller bearings to bind. There should be a gasket (37) on each side of the shim pack. After the bearings are properly adjusted, check the backlash of the bevel gears. The backlash should be 0.005-0.009. If the backlash is less than 0.005, remove a shim (22); or, if the backlash is more than 0.009 add a shim (22).

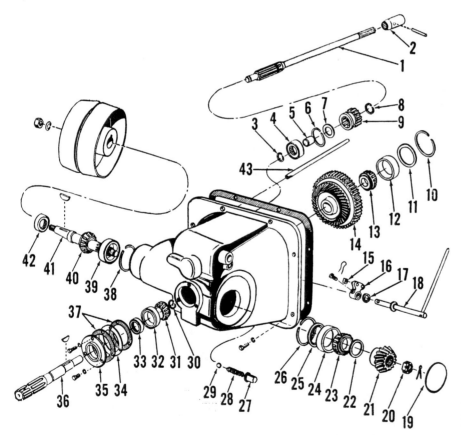

Fig. CS280—Exploded view of model 20 combination belt pulley and power take-off unit. The unit receives its drive from the transmission mainshaft.

1. Drive shaft
2. Drive sleeve
3. Snap ring
4. Bearing
5. Bearing inner race
6. Snap ring
7. Spacer
8. Snap ring
9. Drive gear
10. Snap ring
11. Shims
12. Bearing cup
13. Bearing cone
14. Spur and spiral bevel gear
15. Trunnion
16. Shift finger
17. Cork washer
18. Shift lever
19. Welch plug
20. Nut
21. Pulley drive gear
22. Shims
23. Bearing cone
24. Bearing cup
25. Washer
26. Snap ring
27. Retainer for spring
28. Spring
29. Lock ball
30. Snap ring
31. Bearing cone
32. Bearing cup
33. Oil seal
34. Shims
35. Bearing cap
36. PTO shaft
37. Gasket
38. Snap ring
39. Bearing cup
40. Bearing cone
41. Belt pulley shaft
42. Oil seal
43. Oil tube

BELT PULLEY UNIT

Model 30

The belt pulley unit (Fig. CS283) is driven by a bevel gear (20) which is mounted on the engine clutch shaft. The driving and driven bevel gears are available only in a matched set. To remove the driving bevel gear, remove the clutch shaft as outlined in paragraph 221.

325. OVERHAUL. The belt pulley assembly is composed of two units which can be separated and overhauled separately.

OUTER UNIT. The procedure for removing the unit is evident. To disassemble the outer unit, unlock and remove nut (13—Fig. CS283). Remove pulley and Woodruff key and bump shaft (9) out of housing. The need and procedure for further disassembly is evident.

When reassembling the unit, install the proper thickness spacer washer (15) so that a torque of 2-6 inch pounds is required to turn the pulley when nut (13) is securely tightened. The spacer washers are available in the following thicknesses; 0.136, 0.141, 0.146, 0.148, 0.151, 0.156, 0.158, 0.161 and 0.166.

INNER UNIT. The procedure for removing the inner part of the pulley unit is evident. To disassemble the unit, remove snap ring (1) and shim washers (X). Bump shaft out of housing. The need for further disassembly is evident.

When reassembling the unit, install the proper thickness of shims (X) to remove all shaft end play without pre-loading the bearings.

325A. To install the inner part of the belt pulley unit, proceed as follows: Remove the cover from top of intermediate case. Bolt the inner part of the pulley unit to the intermediate case using the same thickness of shims between the pulley housing and the intermediate case as were originally installed. Mount a dial indicator as shown in Fig. CS284 and check the bevel gear backlash which should be 0.010-0.014. If the backlash is less than 0.010, add a shim between the pulley housing and the intermediate case. The maximum allowable backlash should never exceed 0.020. The backlash is reduced by removing a shim. Shims are available in thicknesses of 0.003 and 0.010.

Models 40-50

The belt pulley unit (Fig. CS286) is driven by the high and low range drive gear which is located in the forward portion of the transmission case. To remove the high and low range drive gear, follow the procedure outlined in paragraph 266.

327. OVERHAUL. To overhaul the belt pulley unit, first remove the unit from tractor and proceed as follows: Remove cotter pin and nut (55—Fig. CS287) and withdraw pulley. Remove cap (40) and save shims (35) for reinstallation. Remove cap (56) and save shims (58) for reinstallation. Using a soft drift, bump shaft (61) out of housing. Remove covers (66) and (47) from housing. Extract snap ring (30), shim washers (31) and bump shaft (41) out of housing. The need and procedure for further disassembly is evident.

After the unit is disassembled, examine all parts and renew any which are excessively worn. Bevel gears (37 and 49) are available only as a matched set. If the bevel gears require renewal, observe the numbers on the old set of bevel gears; and if possible, install a new set of gears having the same numbers. If the new set of gears has the same numbers as the old set, then use the same number and thickness of shims (46) as were originally removed. If, however, the numbers on the new set of gears is 0.001 larger than those on the old set, then use the proper combination of shims (46) so the total thickness of the shim pack will be 0.001 less than the total thickness of the original shim pack. If, on the other hand, the numbers on the new set of gears are smaller, it will be necessary to add the equivalent thickness of shims. Shims (46) are available in thickness of 0.003 and 0.005.

Fig. CS284—Dial Indicator mounted for checking the backlash between the model 30 belt pulley bevel gears.

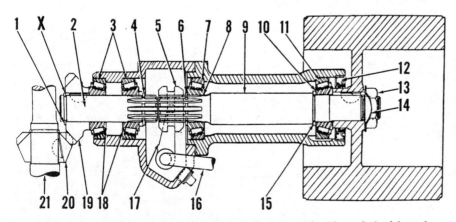

Fig. CS283—Sectional view of model 30 belt pulley assembly. The unit is driven from the bevel gear (20) which is mounted on the engine clutch shaft (21).

X. Shims	6. Snap ring	11. Bearing cup	17. Fork
1. Snap ring	7. Bearing cup	12. Oil seal	18. Bearing cone
2. Inner drive shaft	8. Bearing cone	13. Nut	19. Driven bevel gear
4. Snap ring	9. Outer drive shaft	14. Locking washer	20. Driving bevel gear
5. Clutch collar	10. Bearing cone	15. Shim washer	21. Engine clutch shaft

COCKSHUTT 20-30-40-50 — Paragraph 327

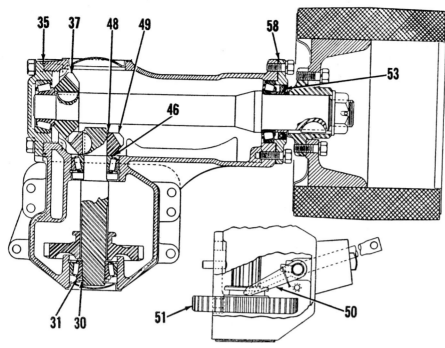

When installing shaft (41) vary the thickness of shim washer pack (31) to provide a slight pre-load for the taper roller bearings. Shim washers (31) are available in thicknesses of 0.1625, 0.1675, 0.1725, 0.1775 and 0.1825.

When installing shaft (61), vary the number of shims (35) to provide a backlash of 0.004-0.006 between the bevel gears, and vary the number of shims (58) to give the pulley shaft an end play of 0.000-0.002. Shims (35) are available in thicknesses of 0.003, 0.005 and 0.010. Shims (58) are available in thicknesses of 0.003, 0.005, 0.010 and 0.020. When installing covers (47 and 66), use Permatex or equivalent to obtain a better seal.

Fig. CS286—Sectional views of models 40 and 50 belt pulley assembly.

30. Snap ring
31. Shim washers
35. Shims
46. Shim washers
48. Snap ring
49. Bevel gear
50. Shift fork
51. Sliding gear
53. Oil seal
58. Shims

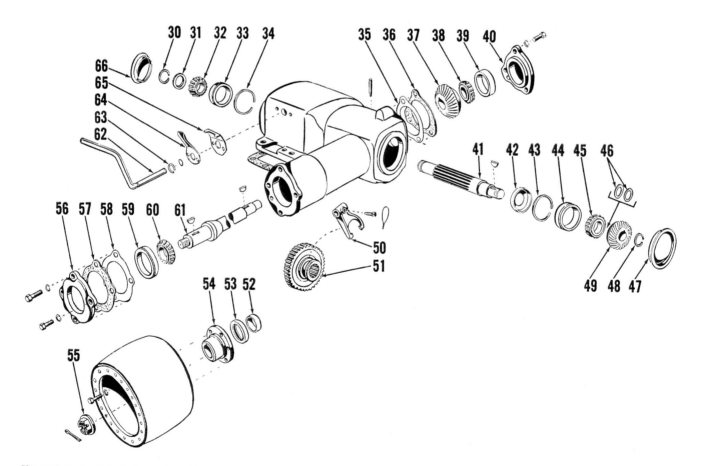

Fig. CS287—Exploded view of models 40 and 50 belt pulley assembly. The unit is driven by the high and low range drive gear which is located in the forward portion of the transmission case.

30. Snap ring
31. Shim washers
32. Bearing cone
33. Bearing cup
34. Snap ring
35. Shims
36. Gasket
37. Bevel gear
38. Bearing cone
39. Bearing cup
40. Adjusting cap
41. Drive shaft
42. Oil retainer
43. Snap ring
44. Bearing cup
45. Bearing cone
46. Shim washers
47. End cover
48. Snap ring
49. Bevel gear
50. Shift fork
51. Sliding gear
52. Bearing spacer
53. Oil seal
54. Hub
55. Nut
56. Outer cap
57. Gasket
58. Shims
59. Bearing cup
60. Bearing cone
61. Pulley shaft
62. Shift lever
63. Snap ring
64. Spring
65. Lever stop

POWER TAKE-OFF UNIT

Model 20

Refer to page CS77.

NON-CONTINUOUS TYPE
Model 30

The non-continuous power take-off unit (Fig. CS290) receives its drive from the rear splined portion of the transmission main shaft. The main shaft can be removed as outlined in paragraph 254.

330. OVERHAUL. The procedure for removing the power take-off unit from rear face of transmission case is evident. To disassemble the unit, proceed as follows: Remove bearing cap (61—Fig. CS290) and extract snap ring (59). Bump the shaft (50) forward and out of housing (52). Remove cap (62) and extract snap ring (74). Bump shaft (71) forward and out of housing. The need and procedure for further disassembly is evident after an examination of the unit.

When reassembling the unit, vary the number and thickness of shims (64), which are available in thicknesses of 0.003, 0.005 and 0.010, to remove all end play from shaft (71) without causing the taper roller bearings to bind.

CONTINUOUS (LIVE) TYPE
Model 30

The continuous (live) power take-off unit receives its drive from a spider and ring assembly which is bolted to the engine clutch cover plate. To visualize the power train, refer to Figs. CS292 and 293.

The hollow shaft and gear unit (16) is splined to spider (2). Power is then transmitted via the hollow shaft and gear (16) to gear (7) which is splined to the front end of the front power take-off drive shaft (9). Power is further transmitted through the front drive shaft and the rear drive shaft (14) to gear (84) which is keyed to the back end of the rear drive shaft. Gear (84) is in constant mesh with gear (21) which is splined to the intermediate drive shaft or clutch shaft (78). Gear (23) is riveted and dowelled on late models to the clutch mounting plate (24) and when the clutch is engaged, the power is transmitted from gear (21), through the multiple disc wet type clutch, to gear (23). Gear (23) is in constant mesh with gear (63) which is splined to the power take-off output shaft (61).

332. OVERHAUL. The occasion for overhauling the complete power take-off system will be infrequent. Usually, any failed or worn part will be so positioned that localized repairs can be accomplished. The subsequent paragraphs will be outlined on the basis of local repairs. If a complete overhaul is

Fig. CS292—Exploded view of the front section of the model 30 live power take-off drive. Spider and ring (2) are bolted to the engine clutch cover.

A. Oil seal
B&C. Engine clutch shaft
1. Drive pin
2. Spider and ring
3. Clutch release bearing
4. Bearing carrier
5. Snap ring
6. Bearing
7. Drive gear
8. Snap ring
9. Front drive shaft
10. Oil seal
11. Oil deflector
12. Bearing
13. Oil retainer
14. Rear drive shaft
15. Snap ring
16. Drive gear and shaft
17. Bearing
18. Oil seal
19. Gasket
20. Bearing cap

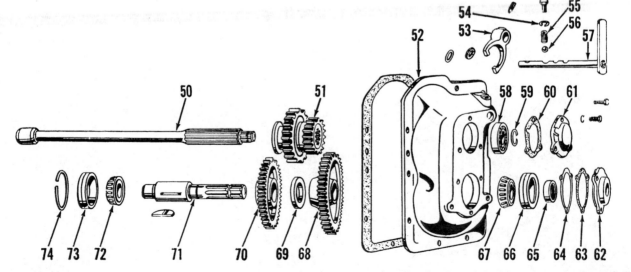

Fig. CS290—Exploded view of model 30 standard, or non-continuous power take-off unit. The unit is driven from the transmission.

50. Drive shaft
51. Sliding gear
52. Case
53. Shift fork
54. Lockwasher
55. Spring
56. Ball
57. Shift lever
58. Bearing
59. Snap ring
60. Gasket
61. Bearing cap
62. Bearing cap
63. Gasket
64. Shims
65. Oil seal
66. Bearing cup
67. Bearing cone
68. Low range gear
69. Spacer
70. High range gear
71. PTO shaft
72. Bearing cone
73. Bearing cup
74. Snap ring

COCKSHUTT 20-30-40-50

Paragraphs 332-335

required, a combination of the appropriate paragraphs can be used.

Early production tractors were factory equipped with a power take-off clutch of the so-called regular duty type, and the unit was not equipped with a braking mechanism. Cockshutt Service Bulletin No. C186-49, dated November 25, 1949 outlined the installation of a heavy duty live power take-off clutch kit No. TFO-10204 on early production tractors. Cockshutt Service Bulletin No. C187-49, dated November 25, 1949 outlined the installation of a brake assembly kit No. TFO-10203 on early production tractors. All subsequent tractors were factory equipped with the heavy duty clutches and braking mechanisms. Replacement parts for the early production, regular duty clutches are no longer available for service; therefore, only the heavy duty type will be considered in this section.

333. RENEW CLUTCH PLATES. To renew the clutch plates (26 and/or 60 —Fig. CS293), first remove the power take-off shaft shield. Remove bearing caps (46) and (47). Unbolt case (41) from the intermediate adapter case (81). Pull the clutch engaging lever (43) rearward and withdraw the rear case (41) as shown in Fig. CS294. The clutch assembly and the output shaft will remain on the adapter case as shown in Fig. CS295. Withdraw the clutch assembly from the adapter case.

334. Using a suitable puller, remove bearing (40—Fig. CS293). Remove the six cap screws retaining the clutch cover bracket (57) to the clutch mounting plate and remove the clutch cover and pressure plate assembly and the clutch plates (26) and (60).

335. Before reassembling the unit, first make certain that bearing (69) is

Fig. CS294—Front view of the removed model 30 live pto case. See Fig. CS293 for legend.

Fig. CS295—Model 30 live power take-off clutch assembly and output shaft. The pto case has been removed as shown in Fig. CS294.

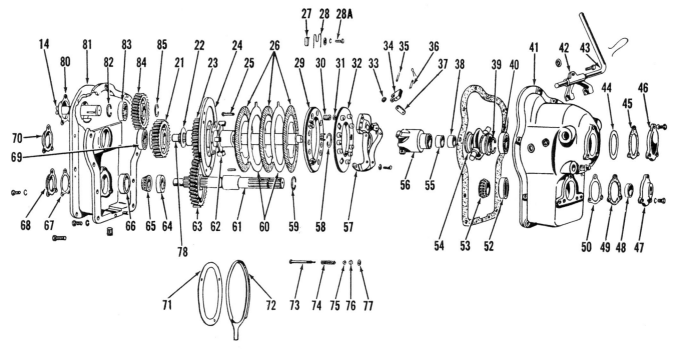

Fig. CS293—Exploded view of model 30 live power take-off rear section. The front section is shown in Fig. CS292.

14. Rear drive shaft	29. Pressure plate	41. Case	53. Bearing cone	64. Spacer	75. Jam nut
21. Intermediate driven gear	30. Pressure spring	42. Fork	54. Bronze bearing	65. Bearing cone	76. Adjusting nut
22. Spacer	31. Washer	43. Shift lever	55. Bushing	66. Bearing cup	77. Shim washer
23. Gear	32. Pressure ring	44. Shims	56. Release sleeve	67. Gasket	80. Cap
24. Mounting plate	33. Roller	45. Gasket	57. Cover plate	68. Cap	81. Intermediate PTO case
25. Rivet	34. Lever	46. Bearing cap	58. Snap ring	69. Bearing	82. Snap ring
26. Facings	35 & 36. Pin	47. Bearing cap	59. Snap ring	70. Cap	83. Bearing
27. Spacer block	37. Link	48. Oil seal	60. Clutch plate	71. Brake plate	84. Intermediate drive gear
28. Return spring	38. Bushing	49. Gasket	61. PTO shaft	72. Stationary plate	85. Snap ring
28A. Cap screw	39. Snap ring	50. Shims	62. Hub	73. Screw	
	40. Bearing	52. Bearing cup	63. Driven gear	74. Spring	

CS-81

pressed all the way on the clutch shaft (78). If the bearing is not completely in place, and the clutch is assembled, one of the lined plates will slip out of position and the clutch will not operate properly. Install the clutch cover bracket assembly making certain that one of the 0.050 thick spacer shims (X—Fig. CS296) is installed under each of the cover bracket mounting feet and that lugs on plates (60) engage and do not bind in grooves of spacers (Y). Securely tighten the six cover retaining cap screws.

335A. Adjust the braking mechanism bumper spring nuts to obtain the dimension (A) shown in Fig. CS297. The distance (A) between the three bumper plates (79) and the shim washers (77) should be 1/32 inch. If the dimension (A) is not 1/32 inch, loosen jam nut (75) and turn adjusting nut (76) until the 1/32 inch clearance is obtained.

336. To reinstall the assembled clutch, place the clutch unit in the adapter case as shown in Fig. CS295. The lug on brake plate (72—Fig. CS296) must be positioned to the right side of the output shaft when viewed from rear. Install case (41—Fig. CS294) making certain that shifter fork (42) engages the release yoke (54—Fig. CS295). Adjust the output shaft bearings as outlined in paragraph 339 and the clutch shaft end play as follows:

337. Bump bearing (40—Fig. CS293) completely in and install enough shims (44) so that rear face of shim pack will be flush with the bearing cap mounting face of case (41). In other words, the clutch shaft (78) should have zero end play without any binding tendency. Shims (44) are available in thicknesses of 0.021, 0.026, 0.030 and 0.034.

338. OVERHAUL OUTPUT SHAFT. To remove the output shaft (61—Fig. CS293), first remove the clutch assembly from the adapter case as outlined in paragraph 333. Withdraw the output shaft 61. If a suitable puller is available, the bearing cup (66) can be renewed at this time. If a puller is not available, it will be necessary to remove the adapter case from the transmission case, remove bearing cover (68) and bump the cup out. The need and procedure for further disassembly is evident. After the output shaft is reinstalled, adjust the taper roller bearing pre-load as follows:

339. Install more than enough of shims (50) and install cap (47). Measure the end play of the output shaft (61). Remove cap (47) and remove a total thickness of shims (50) equal to 0.002 more than the measured end play. This will give the taper roller bearings the desired pre-load of 0.002. For example, if the measured end play was 0.005, then remove 0.007 thickness of shims.

340. COMPLETE CLUTCH. To overhaul the complete clutch, proceed as follows: First remove the clutch plates as outlined in paragraphs 333 and 334. The procedure for disassembling the cover bracket, pressure plate assembly and release sleeve and yoke assembly is evident and the overhaul procedure is conventional.

The pressure springs (30—Fig. CS 293) should have the following specifications.

Spring free length........1.105 inches
Pounds test @ height......
..............175-193 lbs. @ $\frac{63}{64}$ inch.

Renew any spring which is rusted, discolored or does not meet the pressure test specifications.

Examine the pressure plate (29). The friction surface should be smooth and flat. Inspect drive slots in pressure ring (32) for being excessively worn. Inspect pins (36), levers (34), links (37) and rollers (33) for being damaged. Renew any questionable parts.

Assemble the cover bracket, pressure plate assembly and release sleeve and yoke assembly by reversing the disassembly procedure.

341. When installing the three cap screws (28A), spacer blocks (27) and return springs (28), observe the following: Place the spacer blocks (27) in position. Lay the return springs (28) in position and slide the springs inward toward center of clutch. Place the lock washer and flat washer on the cap screws (28A) and start the cap screws loosely. Using two small screw drivers or similar tools hooked into the loops of the return springs, pull the springs outward until the springs are firmly against the cap screws. Tighten the cap screws securely.

342. Using a suitable puller, remove bearing (69). Withdraw gear (21) and the clutch mounting plate (24). Inspect the friction surface on the clutch mounting plate for being smooth and flat. Inspect the rivets which secure gear (23) to the mounting plate. If the rivets are loose, or missing, rerivet the gear to the mounting plate. Inspect the brake plate (71) and renew same if it is damaged.

When assembling the clutch cover bracket and pressure plate assembly to the clutch mounting plate, refer to paragraph 335.

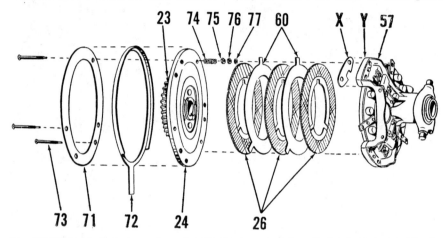

Fig. CS296—Exploded view of model 30 live power take-off clutch plate and brake assembly. Refer to legend under Fig. CS293.

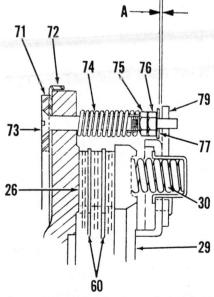

Fig. CS297—Sectional view of model 30 live power take-off clutch braking mechanism, showing the adjustment dimension (A). Models 40 and 50 are similar

29. Pressure plate
30. Pressure spring
60. Clutch plate
71. Brake plate
72. Stationary plate
73. Screw
74. Spring
75. Jam nut
76. Adjusting nut
79. Stop

COCKSHUTT 20-30-40-50

343. RENEW REAR DRIVE SHAFT. To renew the rear pto drive shaft (14—Figs. CS293 or 298), first remove case (41) from tractor and withdraw clutch assembly as in paragraph 333. Shift transmission into third gear and remove intermediate case (81) from rear of transmission case. Remove snap ring (85) and unbolt front cover (80) from intermediate case. Bump shaft out of the intermediate case.

Install shaft (14) by reversing the removal procedure.

344. RENEW FRONT DRIVE SHAFT & GEAR. To renew front pto drive shaft (9—Fig. CS298) and gear (7) proceed as follows: Detach (split) the transmission case from the intermediate case as outlined in paragraph 243. Disconnect cables from starting motor, and on non-Diesel models, disconnect the starter pull rod. Unbolt fuel tank supports from tractor and raise the fuel tank enough to clear the clutch housing cover. Remove starting motor and unbolt clutch housing cover from engine frame. Remove the clutch housing cover. Working through the clutch housing cover opening, remove the bearing cover from front of intermediate case and extract snap ring (5). Using a soft drift, bump the front pto shaft rearward and out of the intermediate case. Withdraw gear (7) from above. The need and procedure for further disassembly is evident.

When reassembling, install oil seal (10) with lip of same facing front of tractor and install deflector (11) with slot of same toward top of tractor.

345. RENEW DRIVING SHAFT AND GEAR. To renew the hollow shaft and gear (16—Fig. CS298), first detach (split) engine frame from the intermediate case as outlined in paragraph 214. Remove the engine clutch release bearing and the release bearing carrier. Remove the clutch shaft cap (20) from front of intermediate case and withdraw the hollow shaft and gear.

To install the shaft and gear, reverse the removal procedure.

346. RENEW DRIVE SPIDER. The procedure for removing the drive spider is evident after removing the clutch as outlined in paragraph 213.

Models 40-50

The continuous (live) power take-off unit receives its drive from a coupling which is attached to the engine flywheel. To visualize the power train, refer to Figs. CS300 and 301.

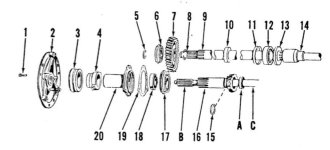

Fig. CS298—Exploded view of the front section of the model 30 live power take-off drive. Refer to Fig. CS292 for legend.

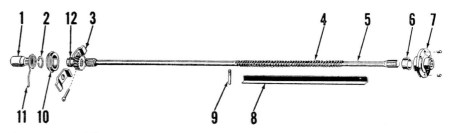

Fig. CS300—Models 40 and 50 accessory unit drive shaft. Coupling (7) is attached to the engine flywheel. Late models have a shifting device for the hydraulic pump drive gear and coupling (12).

1. PTO clutch shaft
2. Snap ring
3. Retainer
4. Spring
5. Drive shaft
6. Bushing (or bearing)
7. Coupling
8. Oiler iron
9. Stud
10. Bearing
11. Oiler spring

Fig. CS300A—Top view of models 40 and 50 final drive case with top cover removed. The accessory drive shaft (5) and coupling (12) can be removed from rear after removing the PTO housing.

Paragraphs 350-352 — COCKSHUTT 20-30-40-50

The accessory drive shaft (5) is splined at the front end to the engine flywheel coupling (7) and at the rear end, to the accessory drive gear and collar (12). Clutch shaft (1) is splined to the accessory drive gear and collar (12) and gear (44) is bolted to the clutch mounting plate (17). When the clutch is engaged, the power is transmitted from shaft (1), through the multiple disc type clutch, to gear (44). Gear (44) is in constant mesh with gear (48) which is splined to the power take-off output shaft (49).

350. OVERHAUL. The occasion for overhauling the complete power take-off system will be infrequent. Usually, any failed or worn part will be so positioned that localized repairs can be accomplished.

The subsequent paragraphs will be outlined on the basis of local repairs. If a complete overhaul is required, a combination of the appropriate paragraphs can be used.

351. CLUTCH PLATES. To renew the clutch plates (15 and/or 41—Fig. CS301), first remove the complete power take-off housing assembly from rear of final drive case. Remove caps (33 and 34). Remove bearing retainer (45) and withdraw gear (48) and shaft (49). Using a soft drift against rear of clutch shaft (1), bump the clutch shaft forward and withdraw the complete clutch assembly.

Remove the six cap screws retaining the clutch cover bracket (12) to the clutch mounting plate and remove the clutch cover and pressure plate assembly and the clutch plates (15) and (41).

351A. When reassembling the unit, be certain to install one 0.050 thick spacer (52) under each of the cover bracket mounting feet and make certain that lugs on plates (15) engage and do not bind in grooves of spacers on cover bracket (12). Tighten the cover bracket retaining cap screws securely. Adjust the bumper spring gap as outlined in paragraph 335A.

Install the assembled clutch unit by reversing the removal procedure and adjust the power take-off shaft bearings as outlined in paragraph 353.

352. OUTPUT SHAFT. To remove output shaft (49), first remove the complete power take-off housing assembly from rear of final drive case. Remove cap (34—Fig. CS301) and bearing retainer (45). Withdraw gear (48) and

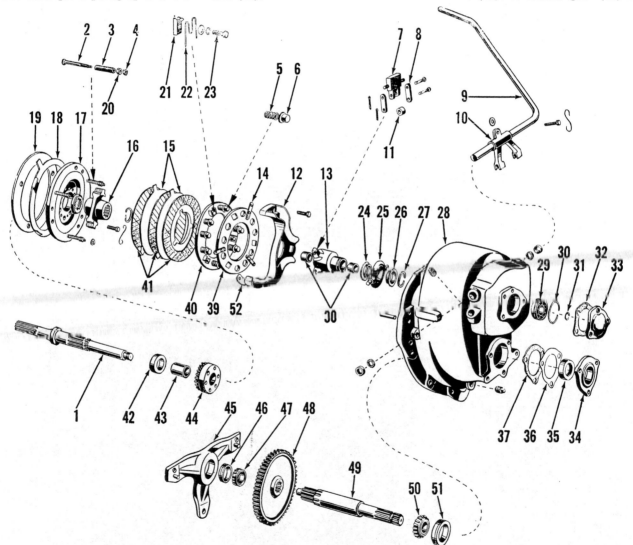

Fig. CS301—Exploded view of models 40 and 50 live power take-off clutch and associated parts.

1. Clutch shaft	10. Fork	18. Stationary plate	27. Snap ring	35. Oil seal	43. Bearing
2. Screw	11. Roller	19. Brake plate	28. Case	36. Shims	44. Drive gear
3. Spring	12. Cover bracket	20. Jam nut	29. Bearing	37. Gasket	45. Bearing retainer
4. Adjusting nut	13. Release sleeve	21. Spacer block	30. Snap ring	38. Bushings	46. Bearing cup
5. Pressure spring	14. Stop	22. Return spring	31. Snap ring	39. Pressure ring	47. Bearing cone
6. Spring cup	15. Clutch plate	24. Bronze bearing	32. Gasket	40. Pressure plate	48. Gear
7. Lever	16. Hub	25. Yoke	33. Cap	41. Facings	49. PTO output shaft
8. Link	17. Mounting plate	26. Bearing ring	34. Cap	42. Collar	50. Bearing cone
9. Shift lever					51. Bearing cup

CS-84

shaft (49). The need for further disassembly is evident.

353. When reinstalling the output shaft, adjust the taper roller bearings by adding or removing shims (36) until the shaft has an end play of 0.003-0.005.

354. COMPLETE CLUTCH. To overhaul clutch, proceed as follows: Remove the clutch plates as outlined in paragraph 351. The procedure for disassembling the cover bracket, pressure plate assembly and release sleeve and yoke assembly is evident and the overhaul procedure is conventional. Check the pressure springs (5—Fig. CS301) against the following specifications.

Spring free length........1.105 inches
Pounds test @ height
      175-193 lbs. @ $\frac{63}{64}$ inches

Examine the pressure plate (40). The friction surface should be smooth and flat. Inspect drive slots in pressure ring (14) for being excessively worn. Inspect pins, levers (7), links (8) and rollers (11) for being damaged. Renew any questionable parts.

Assemble the cover bracket, pressure plate assembly and release sleeve and yoke assembly by reversing the disassembly procedure. When installing the three cap screws (23), spacer blocks (21) and return springs (22), observe the following: Place the spacer blocks (21) in position. Lay the return springs (22) in position and slide the springs inward toward center of clutch. Place the lock washer and flat washer on the cap screws (23) and start the cap screws loosely. Using two small screw drivers or similar tools hooked into the loops of the return springs, pull the springs outward until the springs are firmly against the cap screws. Tighten the cap screws securely.

When assembling the clutch cover bracket and pressure plate assembly to the clutch mounting plate, refer to paragraph 351A.

355. ACCESSORY SHAFT AND/OR GEAR. To renew the accessory drive shaft (5—Fig. CS300) on models without the hydraulic pump drive gear shifting mechanism, proceed as follows: Remove the complete power take-off housing assembly. Remove both bearing retainers (3 — Fig. CS300) and withdraw the accessory shaft (5) and gear (12) from tractor.

Install the shaft and gear by reversing the removal procedure.

On models equipped with the hydraulic pump drive gear shifting mechanism, remove the accessory drive shaft and gear (13—Fig. CS302) as follows: Remove the final drive case top cover and the power take-off unit. Remove the shaft rear bearing retainers, and disengage snap ring (12) from coupling (11). Withdraw drive shaft from rear and gear (13) from above.

356. DRIVE COUPLING. The drive coupling (7—Fig. CS300), which is attached to the engine flywheel, can be renewed after removing the engine clutch as outlined in paragraph 216. The drive coupling contains the pilot bearing (or bushing) for the forward end of the engine clutch shaft. On model 40 tractors prior to serial No. 11238 and model 50 tractors prior to serial No. 1297, a bushing type pilot is used. On later models, a ball type pilot bearing is used. Refer to paragraph 218 for information concerning the change over from the bushing to the ball type.

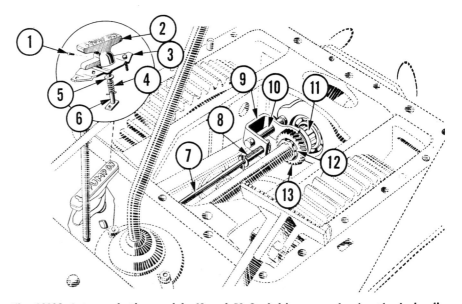

Fig. CS302—Late production models 40 and 50 final drive case, showing the hydraulic pump drive gear shifting mechanism.

1. Set screw
2. Shifter handle
3. Plate
4. Spring
5. Seal
6. Shift lever
7. Shift rod
8. Spring
9. Shifter fork
10. Shift rail
11. Coupling sleeve
12. Snap ring
13. Gear

HYDRAULIC SYSTEM

NOTE: The maintenance of absolute cleanliness of all parts is of utmost importance in the operation and servicing of the hydraulic system. Of equal importance is the avoidance of nicks or burrs on any of the working parts.

LUBRICATION
Models 20-30-40-50

365. It is recommended that the hydraulic system working fluid (S.A.E. 10W oil) be changed once-a-year or every 500 hours of operation. Drain the system while oil is warm. Refill the oil reservoir, start engine and run until hydraulic oil is at normal operating temperature. Operate the lift system several times and check for leaks at all connections. Stop engine and refill reservoir if necessary.

If the system is to be operated in extremely cold temperatures, it is advisable to use S.A.E. 5W oil for the working fluid.

SYSTEM ADJUSTMENTS
Models 20-30-40-50

366. The subsequent paragraphs outline the adjustments which can be performed in the field. The need for a particular adjustment is usually evident after observing the action of the system.

367. RESTRICTOR ELBOW. The rockshaft operating or remote control cylinder on models 40 and 50 and the remote control cylinder on models 20 and 30 are equipped with a restrictor

elbow assembly which can be adjusted to provide smooth lowering of implements. To make the adjustment, start engine and run at the slow idle speed. Loosen the lock nut and turn adjusting screw (1—Fig. CS310 or 311) in slowly until the implement can be lowered smoothly without jerking.

Move the engine throttle lever to the full speed position and check to make certain that control lever does not unlatch before the cylinder completes its stroke and that implement will raise. If the control level unlatches prematurely or if implement will not raise, the restrictor elbow screw has been turned in too far and same should be backed out slightly.

368. CYLINDER STOP. On models 40 and 50, when the work cylinder is used for operating the rockshaft, the cylinder stop (2—Fig. CS311) should be in the nearest groove to the cylinder anchor. On all models, where the cylinder is used for operating trailing implements, the stop (2—Fig. CS310) can be positioned in any groove which will give the desired working depth. A finer control of depth adjustment can be obtained by placing the spacer (3) between the cylinder stop and the cylinder distance piece.

369. YOKE ADJUSTMENTS. Note: These adjustments apply only to models 20 and 30 when using mounted equipment or on model 30 when using a Dowden loader. To make the adjustments, remove the reservoir cover and proceed as follows: Place the main control lever in neutral position; at which time, tip (2A—Fig. CS312) should be in a vertical position. Gently push plunger (3) rearward against ball (1) and turn adjusting screw (4) until the clearance between plunger (3) and the head of screw (4) is 1/16 inch for model 30, 1/32 inch for model 20. The clearance is shown at (A). Adjust cap screw (6) on the yoke (2) so that when the control lever is moved very slightly forward of neutral position, the cam (5) is just leaving the yoke (2).

Move the control lever back until the cam (5—Fig. CS313) is in the shallower notch on the float release yoke (2). Loosen the locknut and turn the screw (7) until it just touches shift plate (8).

370. UNLATCH VALVE. This adjustment applies to models 20, 30, 40 and 50. If the control lever does not maintain its position while lifting a normal load and tends to go back to neutral before the end of the cylinder stroke, remove the reservoir cover and plug (6—Fig. CS314). Add one or more 1/32 inch thick washers (8) until the lever will maintain its position until load is lifted.

371. RELIEF VALVE. This adjustment applies to models 20, 30, 40 and 50. To check and/or adjust the relief valve opening pressure, install a suitable pressure gage in one of the reservoir discharge ports. On models 20 and 30 which are not equipped with a remote control cylinder remove either plug (P—Fig. CS315) and install the gage. On models 20 and 30 which are equipped with a remote control cylinder and all models 40 and 50, it will be necessary to disconnect one of the cylinder lines from the reservoir and install the gage in series between the line and the reservoir.

Operate the system and observe the pressure at which the valve opens. The relief valve opening point can be detected by a very high pitched noise. The gage should show a pressure of 1225-1375 psi. If the gage pressure is not as specified, remove the reservoir cover and plug (39—Fig. CS314) and vary the number of 1/32

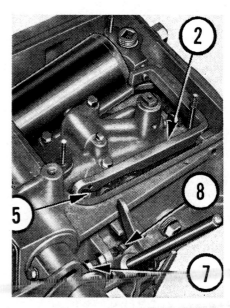

Fig. CS313—Models 20 and 30 float release yoke adjustments.

2. Yoke   7. Adjusting screw
5. Cam   8. Shift plate

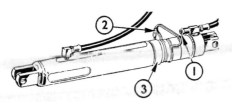

Fig. CS310—Adjustments for a typical remote control cylinder.

1. Adjusting screw 2. Stop 3. Spacer

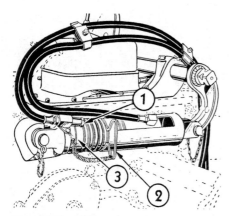

Fig. CS311—Adjustment points for models 40 and 50 rockshaft operating cylinder.

1. Adjusting screw 2. Stop 3. Spacer

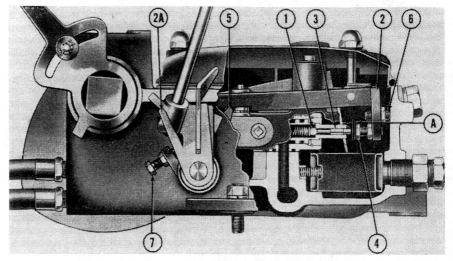

Fig. CS312—Cut-away view of models 20 and 30 hydraulic system reservoir and control valves unit.

1. Ball   2A. Tip   4. Adjusting screw   6. Adjusting screw
2. Yoke   3. Plunger   5. Cam   7. Adjusting screw

COCKSHUTT 20-30-40-50

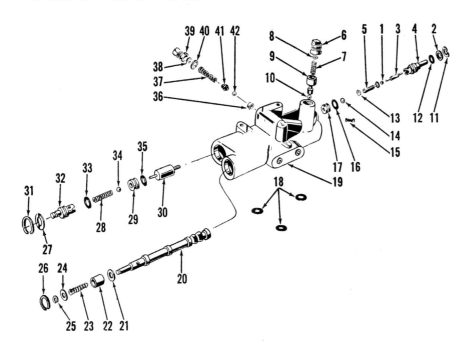

Fig. CS314—Exploded view of models 20 and 30 hydraulic system control valves unit. Models 40 and 50 are similar.

1. Control valve ball
2. Washer
3. Float release plunger
4. Plug
5. Spring
6. Drilled plug
7. Spring
8. Adjusting washer
9. Piston
10. Ball
11. Snap ring
12. Washer
13. Snap ring
14. Ball
15. Drilled plug
16. Washer
17. Valve seat
18. "O" rings
19. Valve body
20. Valve spool
21. Washer
22. Spacer
23. Spring
24. Washer
25. Snap ring
26. Snap ring
27. Washer
28. Spring
29. Valve seat
30. Piston
31. Snap ring
32. Plug
33. Washer
34. Ball
35. Washer
36. Seat
37. Spring
38. Spacer
39. Plug
40. Washer
41. Retainer
42. Ball

Fig. CS315—Rear view of models 20 and 30 hydraulic system reservoir. Remote cylinder hoses can be installed after removing plugs (P).

inch thick washers (38) to obtain the desired opening pressure. If the pressure was too low, add washers; if the pressure was too high, remove washers. Always recheck the pressure reading after making an adjustment.

CONTROL VALVES UNIT
Models 20-30-40-50

375. OVERHAUL. Models 20, 40, 50, and late model 30 tractors are equipped with the control valves unit shown in Fig. CS314. The unit is mounted within the hydraulic system reservoir and can be removed after removing the reservoir cover. Overhaul of the unit consists of completely disassembling, cleaning and renewing any damaged parts.

The 0.9895-0.990 diameter spool (20) has a clearance of 0.0003-0.0018 in the 0.9903-0.9913 diameter spool bore. The 0.9047-0.9052 diameter piston (30) has a clearance of 0.0005-0.002 in the 0.9057-0.9067 diameter piston bore.

When reassembling, renew all "O" ring seals and make certain that snap rings are securely seated. Plugs (6) and (15) have a small hole drilled through them and must be installed in the position shown.

After the unit is installed, adjust the unlatch valve as in paragraph 370 and the relief valve opening pressure as in paragraph 371.

Early Model 30 Pesco Unit

376. OVERHAUL. Overhaul of the early model 30 Pesco hydraulic control valve unit, which is shown in Fig. CS320, consists of completely disassembling, cleaning and renewing any damaged parts.

To disassemble the unit, remove cover plate (18), and extract spacers (13 and 21). Remove spacer (11), spring (10), ball check and guide assembly (9) and ball (8). Remove cover plate (3) and withdraw the control

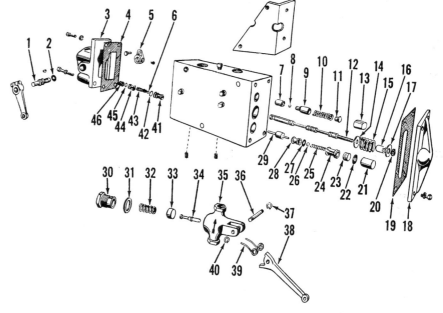

Fig. CS320—Exploded view of early model 30 hydraulic control valves unit.

1. Camshaft
2. Seal ring
3. Cover
4. Gasket
5. Cam
6. Ball
7. Seat
8. Ball
9. Guide
10. Spring
11. Spacer
12. Control spool
13. Spacer
14. Washer
15. Spring
16. Stop
17. Washer
18. Cover
19. Gasket
20. Snap ring
21. Spacer
22. Seal ring
23. Guide
24. Guide
25. Spring
26. Ball
27. Seal ring
28. Seat
29. Shuttle
30. Plug
31. Gasket
32. Spring
33. Poppet
34. Stem
35. By-pass valve
36. Lever pin
37. Snap ring
38. Lever
39. Spring
40. Snap ring
41. Seat
42. Seal ring
43. Spring
44. Guide
45. Seal ring
46. Guide

spool assembly (12). Remove guides (23) and (46) from each end of unit. If guides are difficult to remove, tap ends of housing with a rubber mallet. Note: Guide (46) should be removed first. The remainder of the parts can be removed at this time. Remove all clean out plugs and thoroughly clean the housing passages.

Use Fig. CS320 as a general guide during reassembly and renew all "O" ring seals. When installing shuttle (29), make certain that small end of shuttle is used in seat with large hole.

PUMP

Model 20

380. **R & R AND OVERHAUL.** The vane type pump is mounted on the right front side of the engine and is driven through the governor by the engine timing gear train. The procedure for removing the pump is evident. To disassemble the pump, remove cover (1—Fig. CS322), pressure plate (17) and rotor (18). Remove coupling (10) and snap ring (11). Withdraw shaft assembly. The need and procedure for further disassembly is evident.

After pump is disassembled, thoroughly clean all parts and examine same for being excessively worn. Always renew any questionable parts. A repair kit (Cockshutt part number TO-12641) is available for servicing the pump. The kit includes items (2, 4, 18 and 6).

Model 30 (Non-Diesels)

381. **REMOVE AND REINSTALL.** The gear type hydraulic pump is mounted on the left front side of the engine and is driven through the governor by the engine timing gear train. When removing and reinstalling the pump, do not disturb the governor or pump mounting bracket.

382. **OVERHAUL.** To disassemble the pump, remove coupling (14—Fig. CS323) and unbolt body (1) from cover (16). Remove plug (3) and disassemble the remaining parts. Thoroughly clean all parts and examine same for excessive wear using the specifications which follow:

Gear diameter 1.4495-1.4500
Gear bore in pump body.1.4505-1.4515
Gear shafts diameter....0.6225-0.6235
Shafts bore in bearings..0.625 -0.626
Gear backlash ....0.002 -0.006

When reassembling the pump, use the accompanying illustrations as a guide and be sure to renew all seals.

Model 30 (Diesels)

383. **R & R AND OVERHAUL.** The hydraulic pump on Diesel model 30 tractors is mounted on an adaptor unit which is bolted to the left side of the intermediate case as shown in Fig. CS324. The pump receives its drive from the belt pulley drive pinion which is located on the engine clutch shaft. The procedure for removing the pump is evident and the overhaul procedure is the same as outlined in paragraph 382.

Models 40-50

385. **R & R AND OVERHAUL.** As shown in Fig. CS326, the hydraulic pump is mounted on the bottom side of the final drive case cover. The unit is driven by gear (13—Fig. CS328).

Fig. CS324—Diesel model 30 hydraulic pump installation. The unit is driven by a pinion on the engine clutch shaft.

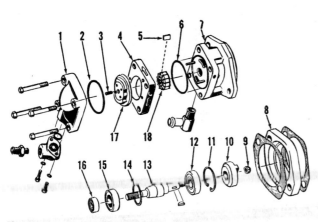

1. Cover
2. "O" ring
3. Spring
4. Ring
5. Vane
6. "O" ring
7. Body
8. Spacer
9. Nut
10. Coupling
11. Snap ring
12. Bearing
13. Drive shaft
14. Spacer
15. Seal
16. Bearing

Fig. CS322—Exploded view of model 20 hydraulic pump. The pump can be serviced with Cockshutt repair kit number TO-12641.

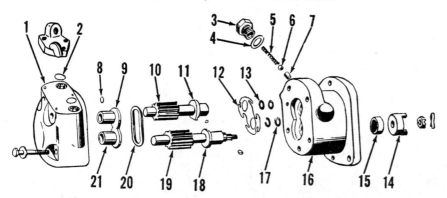

Fig. CS323—Exploded view of late production model 30 hydraulic pump. Early production pumps were similarly constructed except items (3, 4, 5, 6 and 7) are not used.

1. Pump body
2. Packing ring
3. Plug
4. Gasket
5. Spring
6. Ball
7. Seat
8. Seal ring
9. Bearing
10. Idler gear and shaft
11. Bearing
12. Spring
13. Seal rings
14. Coupling
15. Seal
16. Cover
17. Fiber washer
18. Bearing
19. Drive gear and shaft
20. Seal ring
21. Bearing

Fig. CS326—The hydraulic pump on models 40 and 50 is mounted on the bottom side of the final drive housing cover. The unit is driven by a gear on the accessory drive shaft.

COCKSHUTT 20-30-40-50

To remove pump, first remove the final drive case cover; then, remove the reservoir cover and strainer screen. Remove the pump retaining cap screws and withdraw the pump.

Refer to paragraph 382 for overhaul notes and specifications.

PUMP ADAPTOR

Model 30 (Diesel)

386. R & R AND OVERHAUL. To remove the hydraulic pump adaptor, first remove the pump; then unbolt the adaptor from the intermediate case. The disassembly procedure is evident after an examination of the unit and reference to Fig. CS329.

When reassembling the unit, use the proper thickness spacer shim (22) to remove all end play from the shaft. Spacer shims are available in various thicknesses ranging from 0.118 to 0.166.

When reinstalling the adaptor unit, push the unit into the intermediate case as far as it will go, leaving out the shims (18). Using a feeler gage, measure the clearance gap between the adaptor housing flange and the intermediate case. Withdraw the unit and insert shims (18) of a total thickness equal to 0.010 more than the feeler gage measurement.

10. Snap ring
11. Shaft
12. Bearing cup
13. Bearing cup
14. Woodruff key
15. Bevel gear
16. Snap ring
17. Bearing cone
18. Shims
19. Housing
20. Gasket
21. Bearing cone
22. Adjusting shim

ACCESSORY DRIVE COLLAR AND/OR GEAR

Models 40-50

387. REMOVE AND REINSTALL. To remove the accessory drive sleeve and/or hydraulic pump drive gear (11 and 13—Fig. CS328), first remove the final drive case rear cover or power take-off unit and proceed as follows: On models with a pump

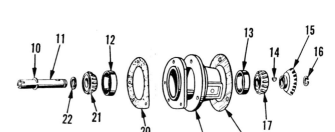

Fig. CS329—Diesel model 30 hydraulic pump adaptor. Gear (15) meshes with the belt pulley drive pinion on the engine clutch shaft.

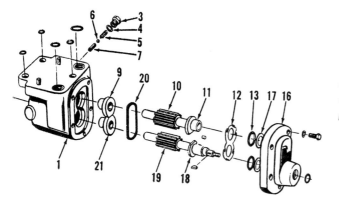

Fig. CS327—Models 40 and 50 hydraulic pump. Refer to legend for Fig. CS323.

shifting mechanism, remove the final drive case top cover and disengage snap ring (12) from coupling (11). On all models, remove cap screws (P—Fig. CS330), bearing retainers (R) and pull drive shaft assembly rearward and out of final drive case. The pump drive gear and/or coupling can be renewed at this time. On early models, without the pump drive gear shifting mechanism, the gear and coupling is integral.

WORK CYLINDERS

Models 20-30-40-50

390. The procedure for disassembling and reassembling the various types of work cylinders is evident after an examination of the unit and reference to the accompanying illustrations. Normal overhaul includes cleaning, and renewing seals and any other questionable parts.

Fig. CS330—Rear view of models 40 and 50 final drive case with PTO housing removed.

R. Bearing retainer
P. Cap screw
T. Spring
S. Oiler angle iron

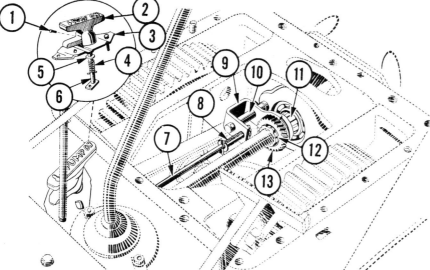

Fig. CS328—Late production models 40 and 50 final drive case, showing the hydraulic pump drive gear shifting mechanism.

1. Set screw
2. Shifter handle
3. Plate
4. Spring
5. Seal
6. Shift lever
7. Shift rod
8. Spring
9. Shifter fork
10. Shift rail
11. Coupling sleeve
12. Snap ring
13. Gear

Paragraph 390   COCKSHUTT 20-30-40-50

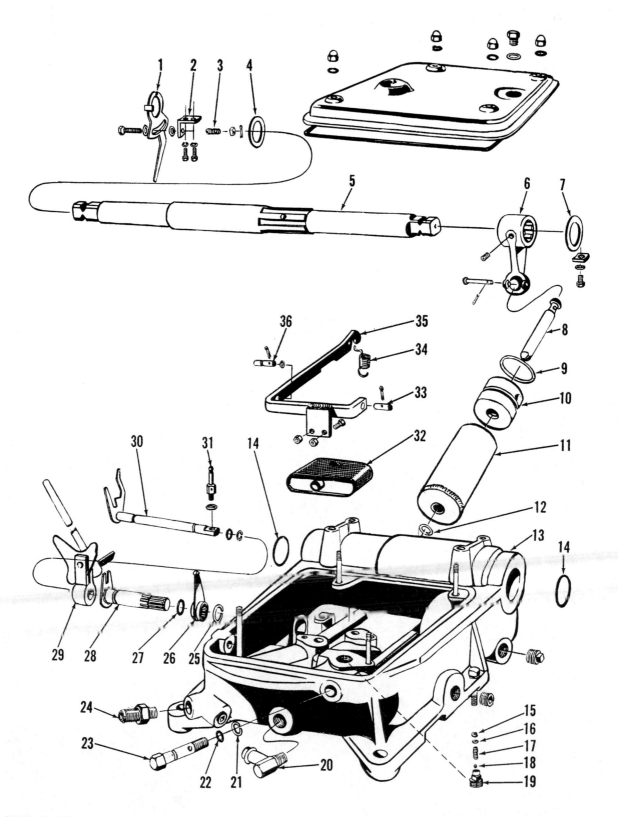

Fig. CS332—Models 20 and 30 hydraulic system reservoir, rockshaft operating cylinder and float release mechanism. The control valves unit shown in Fig. CS314 is mounted within the reservoir.

1. Depth control lever
2. Adjusting bracket sector
3. Spring
4. Washer
5. Rockshaft
6. Rockshaft arm
7. Washer
8. Piston rod
9. "O" ring
10. Piston
11. Cylinder
12. "O" ring
13. Reservoir
14. "O" Ring
15. Retainer ring
16. Washer
17. Thermal relief valve spring
18. Ball
19. Thermal relief valve body
20. Suction elbow
22. "O" ring
23. Cylinder anchor bolt
25. Snap ring
26. Float release cam
27. "O" ring
28. Lock arm and sleeve assembly
29. Shift lever arm
30. Spindle and lock arm assembly
31. Operating pin
32. Filter screen
33. Pivot pin
34. Spring
35. Yoke assembly
36. Pivot pin

COCKSHUTT 20-30-40-50   Paragraph 390

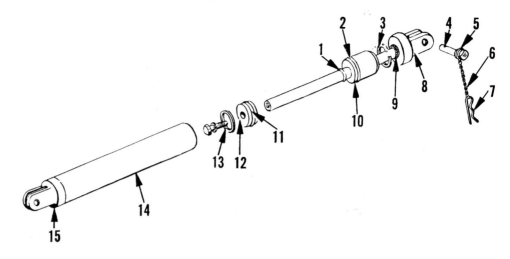

Fig. CS334—Exploded view of an early model 30 hydraulic jack.

1. "O" ring
2. "O" ring
3. Retainer ring
4. Cylinder pin
5. Spring clip
6. Chain
7. Spring clip
8. Shaft and clevis
9. Oil seal
10. Bearing
11. "O" ring
12. Piston
13. Back-up ring
14. Cylinder
15. Clevis

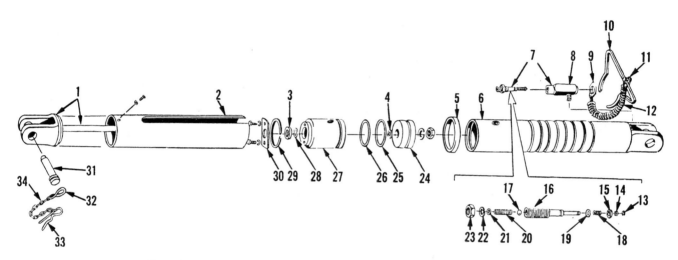

Fig. CS335—Typical hydraulic cylinder as used on models 20, 30, 40 and 50 after 1952.

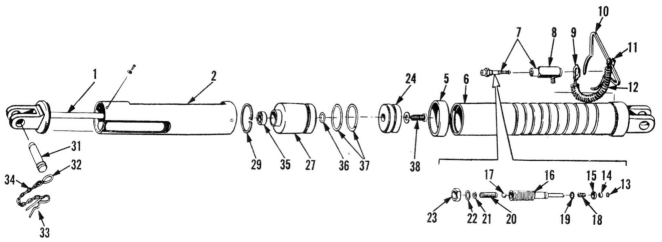

Fig. CS336—Typical hydraulic cylinder as used on models 20, 30, 40 and 50 prior to 1953.

1. Rod and yoke assembly
2. Cylinder sleeve
3. Oil seal
4. Copper washer
5. Cylinder ring
6. Barrel and yoke assembly
7. Restrictor elbow assy.
8. Restrictor elbow body
9. Hose clip
10. Cylinder adjustment stop
11. Adjustment clip
12. Stop spring and chain
13. Snap ring
14. Retainer
15. Restrictor valve
16. Adjusting screw
17. Ball
18. Restrictor spring
19. "O" ring
20. Spring
21. Washer
22. Retainer ring
23. Jam nut
24. Piston
25. "O" ring
26. "O" ring
27. Bearing
28. "O" ring
29. Snap ring
30. Retainer plate
31. Pin
32. Spring clip
33. Spring clip
34. Chain
35. Oil seal
36. "O" ring
37. "O" ring
38. Allen screw

CS-91

Sectional View

COCKSHUTT 20-30-40-50

Cockshutt Model 40. Model 50 is similar.

NOTES

NOTES

NOTES

NOTES

NOTES